ELEGANCE IN SCIENCE

Ian Glynn is Emeritus Professor of Physiology, University of Cambridge and a Fellow of Trinity College, Cambridge. His work on the 'sodium pump' (the molecular machine that keeps the brain batteries charged) led to his election to the Royal Society. He has also served as a member of the Medical Research Council and the Agricultural and Food Research Council.

THE BEAUTY OF SIMPLICITY

IAN GLYNN

OXFORD
UNIVERSITY PRESS

Great Clarendon Street, Oxford, OX2 6DP,
United Kingdom

Oxford University Press is a department of the University of Oxford.
It furthers the University's objective of excellence in research, scholarship,
and education by publishing worldwide. Oxford is a registered trade mark of
Oxford University Press in the UK and in certain other countries

First Edition published in 2010
First published in paperback 2013

Impression: 1

British Library Cataloguing in Publication Data
Data available

Library of Congress Cataloging in Publication Data
Data available

ISBN 978–0–19–957862–7
ISBN 978–0–19–966881–6

Printed in Great Britain on acid-free paper
by Clays Ltd, St Ives plc

CONTENTS

ILLUSTRATIONS

Although every effort has been made to trace and contact copyright holders prior to publication this has not been possible in every case. If notified, the publisher will be pleased to rectify any omissions at the earliest opportunity.

PREFACE

Many years ago I was asked by an undergraduate science society in Cambridge to give a talk about my own research. My colleagues and I had, as it happens, just got some very interesting experimental results, but we were not yet sure that those results, and our interpretation of them, were valid. To talk about work that might later prove to be wrong would be rash. On the other hand, to talk about our older experiments while we were preoccupied with thinking about our more recent ones did not seem very inviting. In this situation, I suggested that I talk about a subject that had fascinated me since my schooldays: the nature and attractiveness of elegance in science. Mathematicians, especially pure mathematicians, are, of course, well known to get excited about elegance. But though scientists talk less about it, elegant theories and elegant experiments do give great pleasure in a wide variety of scientific fields—and not only to the originators. Anyway, my suggestion was accepted, the talk was given, and I wondered at that time whether it could usefully give birth to a book.

Although more than twenty years have passed since then, it is only in the last few years that I have had time to think seriously about the book. There are two obvious difficulties. Firstly, to appreciate the elegance of a theory or an experiment, the reader needs to be aware of the state of play in the relevant field at the time that that theory or experiment was created; and elegant theories and experiments have been around for a very long time.

Inevitably, then, each topic has to be seen against its historical background. Fortunately, the fascinating character of so many of the scientists involved will, I hope, make the necessary historical digressions something of a bonus.

Secondly, because elegance is found in all branches of science, from the most mathematical to clinical neurology, the state of play needs to be presented in a way that is intelligible to readers unfamiliar with the particular field. In most of the book I have, therefore, excluded topics that are too sophisticated or too mathematical; and where they are not excluded I have either made their character clear in the text or relegated discussion to an appendix. Many elegant theories and experiments are, of course, famous, but in choosing topics I have tried to show that such theories and experiments can also be found in areas where the adjective elegant is not often used. And because the book is not a catalogue, and I want it to be fairly short, it is likely that many readers will find favourite examples not mentioned.

I have been greatly helped in writing the book by my colleagues, my friends, and my family. Graeme Mitchison, mathematician, neuroscientist, molecular biologist, and physicist, gave helpful advice and criticism, whatever the topic. For advice on mathematics and the physical sciences, I am grateful to Béla Bollobas, Anson Cheung, Martin Cowley, John Davidson, Marie Farge, Tim Gowers, Hugh Hunt, David Khmelnitskii, Peter Littlewood, Piero Migliorato, Chris Morlcy, Hugh Osborn, Malcolm Perry, Martin Rees, Gordon Squires, and Peter Swinnerton-Dyer. For advice on the biological sciences and medicine, I am grateful to Horace Barlow, Michael Berridge, Andrew Crawford, Doug Fearon, Andrew Huxley, Aaron Klug, Sachiko Kusukawa, John Mollon, Michael Neuberger, and Nigel Unwin. Different members of my family, spread over three generations, have been able to provide expert criticism of my treatment of some topics, and have helped me to assess the response of 'the intelligent layman' to my treatment of others. In particular I am grateful to my wife for her endless patience and skill in suggesting, or commenting on, possible improvements of the text. Finally I must thank my agent, Felicity Bryan, and my editor, Latha Menon together with her colleagues at the Oxford University Press, for their encouragement, advice, and understanding during the successive stages of the book's production.

1

THE MEANING OF ELEGANCE

The dictionary definitions of elegant—graceful, tasteful, of refined luxury—are useless here. 'Elegant economy' would be nearer the mark, but that phrase has been appropriated by Mrs Gaskell to describe the style of life in *Cranford*—which is not what I have in mind. So let me try to explain what I mean by 'elegance in science' by considering three different ways of solving an utterly trivial mathematical problem—a problem of no other use, of no profundity, and no importance. It is this.

Consider a grid forming eight rows of eight squares, so that there are sixty-four squares in all (upper part of Figure 1). Let us suppose that we have thirty-two dominoes each the size of two squares. Obviously, using all the dominoes we can cover all the squares—and there will be many different ways of doing this. Now suppose we remove two squares, taking them from opposite corners of the grid, and we throw away one of the dominoes. The question is: *Can we cover the remaining sixty-two squares with the thirty-one remaining dominoes?*

The total area covered by the thirty-one dominoes is the same as the total area of the sixty-two squares, so the task is not obviously impossible. How can we tell whether it is possible? One way of proceeding would be by systematic trial and error. That would be tedious and inelegant, and the only reason you wouldn't go crazy doing it is that you would need to be crazy already to start.

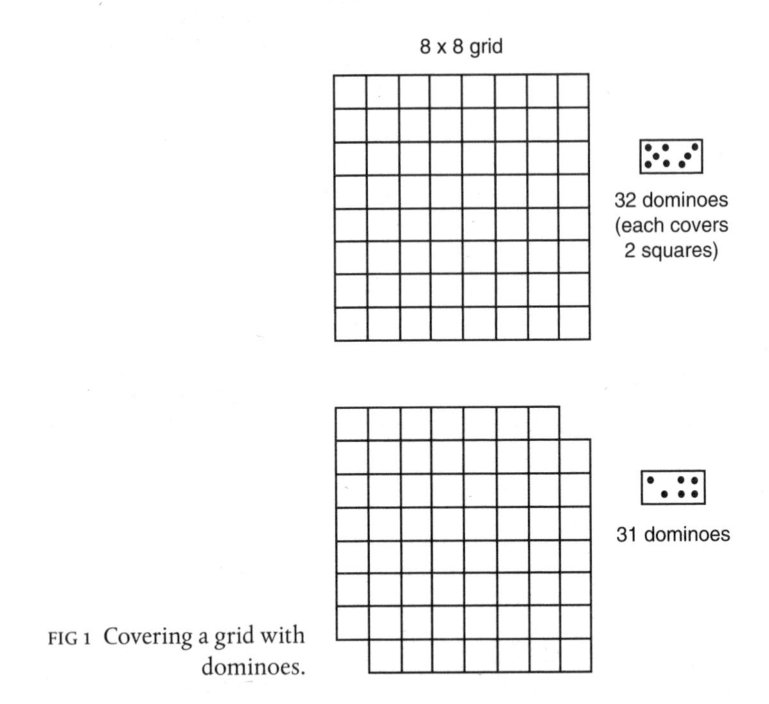

FIG 1 Covering a grid with dominoes.

A second and better method is to assume that the task *is* possible and to see where that assumption leads. Consider the top row of squares. Because of the loss of the corner square there are only seven squares to cover. Any domino laid horizontally in that row will cover two squares, so that however many horizontal dominoes we put in that row they will cover an even number of squares. That will leave an odd number of squares to be filled by vertical dominoes, so an odd number of vertical dominoes will hang down into the second row. That, in turn, means that there is an odd number of squares to be filled in the second row, and the same argument can be repeated. And so we go on until we come to the last row. An odd number of vertical dominoes will hang down into it from the preceding row, and the corner square was missing anyway so there is an even number of squares to be filled by horizontal dominoes. Nothing obviously impossible so far.

But consider the total number of vertical dominoes. We have seven rows, each containing an odd number of hanging vertical dominoes. Seven times an odd number is itself an odd number, so the total number of vertical dominoes must be odd.

Now, if instead of starting with the *top row* of squares and working downwards we start with the *left-hand row* and work to the right, a similar argument will prove that the total number of *horizontal* dominoes must be odd. If there is an odd number of vertical dominoes and an odd number of horizontal dominoes the total number of dominoes must be even. But this can't be right since we know that thirty-one dominoes are needed to cover all the squares. What has gone wrong? Well, the only step in the argument that is uncertain is the initial assumption that the task set is possible. If that assumption leads to two contradictory conclusions it must be wrong. The task must be impossible.

This argument is perfectly sound—even ingenious—and it gives the correct answer to the problem; but it's the sort of argument that leaves us a bit uneasy until we have gone through it a couple of times to make sure there are no snags. So let's come to the third solution.

Imagine that our original grid is a chessboard with alternate black and white squares. Any domino properly placed on the board will inevitably cover one black and one white square. But when we removed a pair of squares from opposite corners of the board they would necessarily have been either both black or both white. It is immediately obvious, then, that we can't cover the board with thirty-one dominoes each of which must cover one black and one white square. That is a very elegant proof indeed. Once one has understood the argument, it can be seen at a glance and one has no doubts about its validity. It is simple, ingenious, concise, and persuasive; it has an unexpected quality; and it is also very satisfying.

It would of course be more satisfying if the problem were not so overwhelmingly trivial. So let's look at some more important problems.

The 'theorem of Pythagoras' has been around for two and a half millennia, though it is not clear that Pythagoras either discovered it or proved it, and it seems to have been known by the Babylonians earlier.[1] It says that in any right-angled triangle the square on the hypotenuse—the side opposite the right angle—is equal in area to the sum of the squares on the other two sides (see Figure 2). It has been suggested that Pythagoras may have become interested in the theorem when he visited Egypt, as the Egyptians were aware that a triangle whose three sides are, respectively, 3, 4, and 5 units long is right-angled; and of course $3^2 + 4^2$ is equal to 5^2. It may also be relevant that the theorem is obviously true for the 'half-square' triangular tiles arranged in the sort of pattern shown in Figure 3, which was a common pattern on Egyptian walls and floors. But '3,4,5 triangles' and 'half-square triangles' are special cases, and Pythagoras' theorem applies to all right-angled triangles.

How, when, and where the theorem was first proved is not known but, three centuries after Pythagoras, Euclid provided a proof that has remained the standard proof ever since. It involves a slightly complicated construction (see Figure 4) and the application of theorems about congruent triangles, and about the relation between the areas of triangles and rectangles that are 'on the same base and between the same parallels'—to use the traditional jargon. For those of us who find it difficult to remember the complicated construction or the relevant theorems, there is another standard

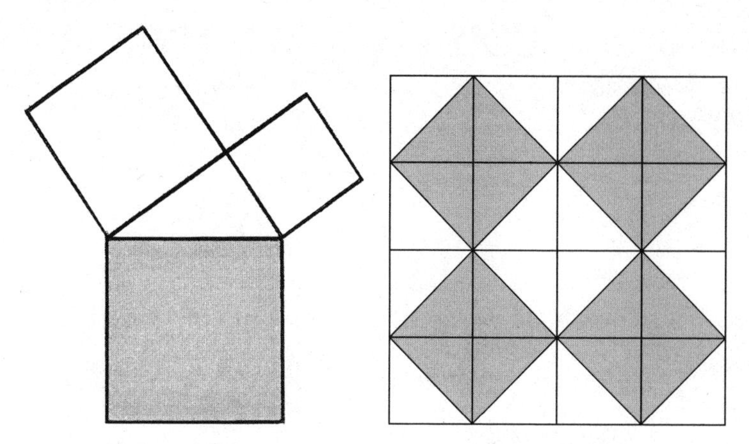

FIG 2 Pythagoras' theorem.

FIG 3 Egyptian tile pattern.

proof with a simpler construction and less geometrical argument but more algebra. Both of these proofs—as well as umpteen others (including one invented by James Garfield, the twentieth President of the United States)—are entirely convincing, but they mostly lack the 'wow' factor. There is, though, a much simpler proof, whose origin is obscure but which was certainly known in the nineteenth century. It is this:

Take four identical right-angled triangles, arrange them as shown in the left half of Figure 5, and draw a frame round them. The area enclosed by the frame is equal to the area occupied by the four triangles plus the area of the central square, which is the square on the hypotenuse. Now rearrange the triangles within the same frame, as shown in the right half of Figure 5. The area enclosed by the frame is now equal to the area occupied by the four triangles plus the areas of two squares, which are the squares on the other two sides of the triangle. Since the area enclosed by the frame and the area occupied by the four triangles remain the

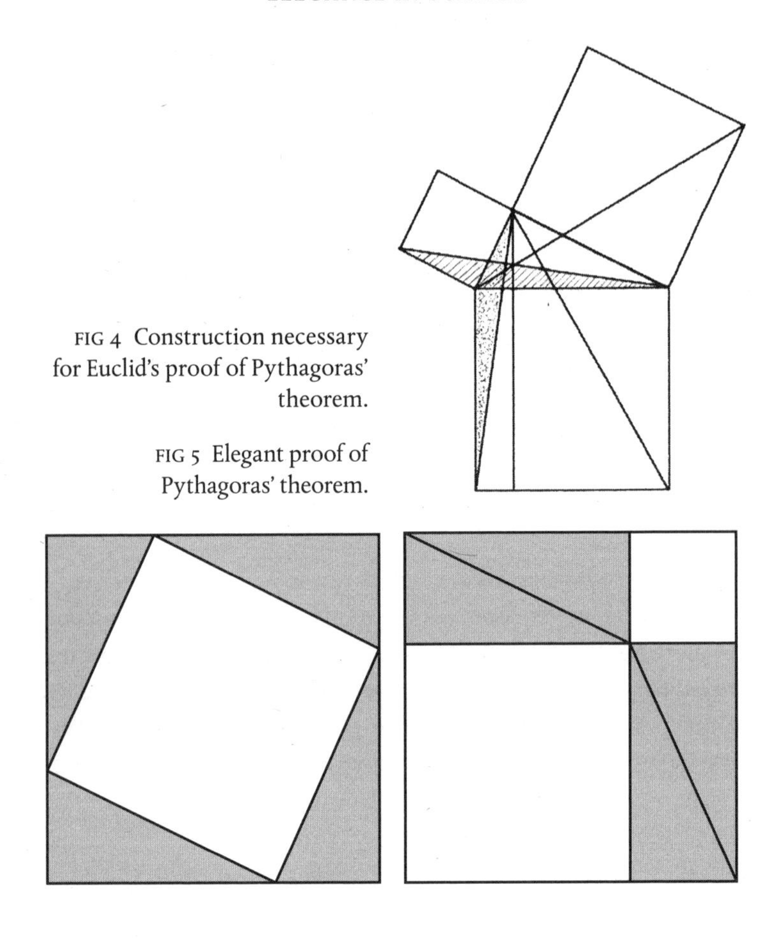

FIG 4 Construction necessary for Euclid's proof of Pythagoras' theorem.

FIG 5 Elegant proof of Pythagoras' theorem.

same, it follows that the square on the hypotenuse must be equal to the sum of the squares on the other two sides. 'Wow!'

A second, important, mathematical problem was solved by Archimedes, who lived in Syracuse in the third century BC. He

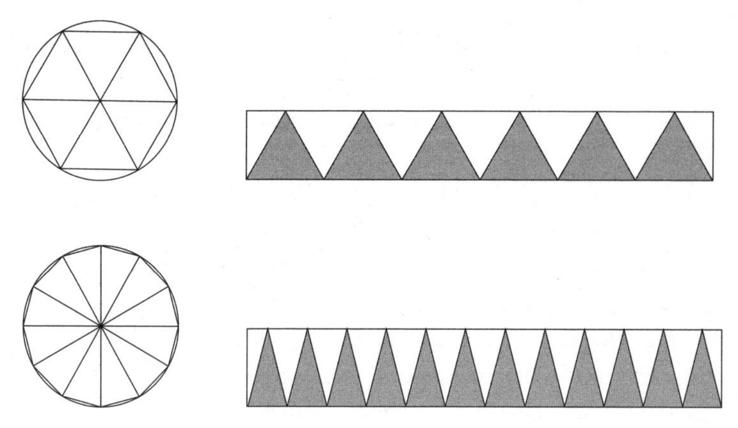

FIG 6 Archimedes' proof that the area of a circle is equal to πr^2.

was the first person to show that the area of a circle is equal to πr^2, where r is the radius and π (he did not actually call it π—the Greek letter pi) is the ratio of the circumference of any circle to its diameter. The proof is not only highly elegant but it foreshadows methods that would be used in the development of the integral calculus two millennia later. The principal of the method—illustrated in Figure 6—is to fit into the circle a succession of regular polygons with an increasing number of sides. As the number of sides increases, the area of the polygon gets closer and closer to the area of the circle; the circumference of the polygon gets closer and closer to the circumference of the circle; and the height of the triangles that make up each polygon gets closer and closer to the radius of the circle. By rearranging the triangles making up each polygon it is easy to see that, as the number of sides is increased, the area of the polygon also gets closer and closer to *half* the area of a rectangle whose width is equal to the radius of the circle and whose length is equal to the circumference of the circle. Since the circumference is $2\pi r$, the area of this rectangle must be $2\pi r^2$, so half the area must be πr^2. It follows that as the number of sides to the polygon increases its area will get closer and closer to πr^2. That must therefore be the area of the circle.

When I showed this proof to my 12-year-old granddaughter, who had just been learning how to calculate the area of a circle, her comment was: 'That's real cool!' So perhaps for her generation being 'real cool' should be added to the criteria for deciding what is elegant in science. But as far as I know, despite the simplicity and venerable age of Archimedes' proof, it is never taught to young children in school. Which is a pity.

Oddly, although Archimedes thought of himself as a mathematician, and asked that his tombstone should commemorate his mathematical work, most of us are familiar with his name either because of his 'principle', which we all learnt at school, or because of the extraordinary story of his detection that a new crown made for the king of Syracuse was not made of pure gold. His principle—that when a solid is wholly or partly immersed in a liquid, the upward force of the liquid on the solid is equal to the weight of liquid displaced—was a profound discovery that took the mystery out of buoyancy. His detective work did not involve his principle though it did involve the displacement of water. The king, who was a friend of Archimedes, was suspicious that in making the crown his jeweller might have kept some of the gold, replacing it with the same weight of silver. Archimedes, so the story goes, was puzzling over this problem as he was about to bathe. Stepping into a full bath, he noticed the water running over, and immediately realized that if he measured the amount of water displaced by the crown, he would know its volume. If he also measured the weight of the crown he could calculate its density, which would tell him whether the gold in the crown had been mixed with the less dense silver. So excited was he that he is said to have dashed naked into the street calling, 'Eureka! Eureka!'—I've found it! I've found it! Whether this now hackneyed story is true is uncertain: we know it from the writings of the architect Vitruvius in the first century BC.

Archimedes, by the way, tended to make memorable remarks after producing ingenious solutions to mechanical problems. Although levers had been used since ancient times, he seems to have been the first person to realize that the ratio between the pull applied to a lever and the force exerted on the load was equal to the ratio between two distances—the distance from the pivot to the load and the distance from the pivot to the point at which the pull is applied. 'Give me a place to stand,' he is reputed to have said, 'and I will move the Earth!'—an exaggerated claim, of course, since he would also need a stand for the pivot, not to mention an exceedingly long pole. But Archimedes' ideas were generally as practical as they were elegant. By turning a cylindrical screw inside a close fitting pipe (see Figure 7) he was able to pump water uphill, whether for irrigation or to empty the flooded hold of a Greek ship.

It was, incidentally, the failure of a modern pump using an 'Archimedean screw' that led to the release of 120 million litres of filtered but otherwise untreated sewage into the Firth of Forth near Edinburgh in April 2007.[2]

The five problems so elegantly solved by Archimedes—How do you find the area of a circle? What determines whether an object floats? How can you tell if your supposedly gold crown is adulterated with silver? How do you arrange a lever if you want to lift a heavy weight with a weak pull? And how do you pump water uphill?—were all problems known to exist. And that is of course

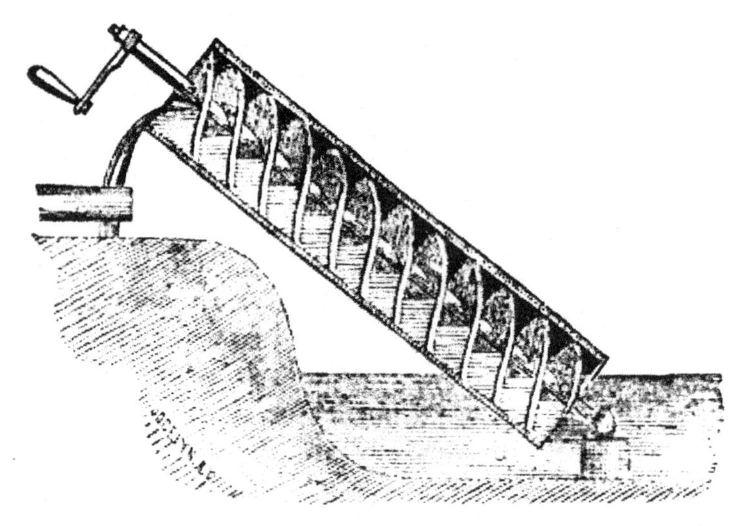

FIG 7 An Archimedean screw used for pumping water.

the usual sequence of events. Rarely, though, an elegant argument reveals the existence of an entirely unsuspected problem. A splendid example of this is what is generally known as Olbers' paradox, though others had been aware of it earlier.[3] Heinrich Olbers was a flourishing physician in Bremen, in the late eighteenth and early nineteenth centuries, with a passion for mathematics and astronomy. At the age of 16, while still at school, he startled everyone by predicting the time of a solar eclipse; at Göttingen he studied maths and physics as well as medicine, and later in life he discovered asteroids and comets and worked out better ways of calculating the paths of comets. When he was 64, he retired from his medical practice, and in the same year he published his paradox. So what is it?

All of us have seen the night sky all our lives and have admired the stars, and taken for granted that in between the stars the sky

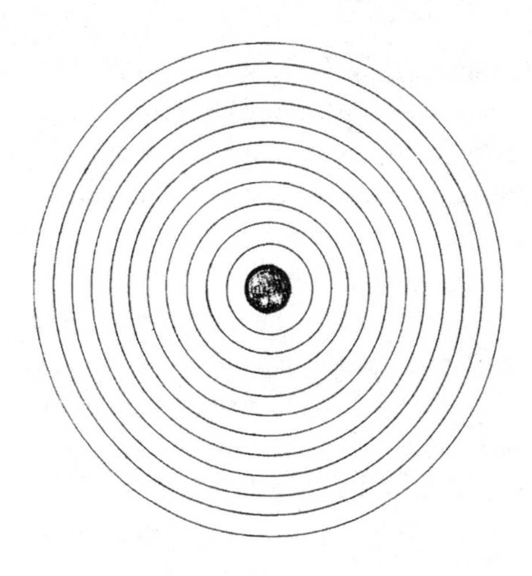

FIG 8 Olbers' paradox.

is black. Olbers' achievement was to realize that to ask why the night sky is black is a worthwhile question.

He supposed that the universe was infinite and roughly uniform in all directions. Consider the light reaching us from the stars in the sky. With ourselves as the centre, think of concentric shells of sky of some arbitrary thickness but all of the same thickness—rather like the layers of an onion (see Figure 8). Now consider two shells, one twice as far from us as the other, and compare the amount of light we are receiving from each. As Kepler had realized in 1604, the amount of light falling on a surface from a point source of light varies inversely as the square of the distance of the surface from the source. (You can see that this must be so by considering a candle in the middle of a hollow sphere and asking how the light falling on a unit area of the inner surface of the sphere varies with the radius of the sphere. Whatever the radius, the total amount of light falling on the inner surface remains the same. Doubling the radius will increase the area of the surface four-fold, so the light falling on a unit area of surface will be one quarter of what it was.) Now, returning to the amount of light we receive from the stars in the two shells, light from individual stars in the further shell will, on average, have to come twice as far, so by the inverse square law, it will be a quarter as intense. But the volume occupied by the further shell will be four times the volume of the nearer shell. (This follows because the shells are of uniform thickness, and the area goes up with the square of the radius.) If the universe is roughly uniform, the further shell will contain four times as many stars, so the effect of the inverse square law will be cancelled out by the increased number of stars. The light reaching us from the further shell, where it is not blocked out by intervening stars, will therefore be of the same average intensity as the light reaching us from the nearer shell. And this will be true for all the other shells. Arguing in this way, you can see that all areas of sky ought to appear extremely bright since the space between nearer stars will allow us to see light from further stars. In other words, in each direction we should be seeing light from stars, and one's intuitive feeling that stars too far away would be too faint is not right because there are more of them.

What follows from the blackness of the sky, then, is that we must be seeing light from only a finite number of stars—either because there is only a finite number, or because, for whatever reason, light from further stars doesn't reach us. Olbers himself thought that the explanation was that space wasn't absolutely transparent because of the presence of interstellar matter. We now believe that the darkness is explained mainly by the fact that there is not an unlimited number of galaxies, and partly by the expansion of the universe. (Because the universe is expanding uniformly, other galaxies are receding from us at speeds proportional to their distances from us. Light from further galaxies will, therefore, seem to us to have longer wavelengths—the so-called 'red shift'—and energy from them will reach us at a lower rate.) But the elegance of Olbers' paradox is that such far-reaching conclusions can be derived from observations that have been available since people first looked at the sky at night, and from theories—elementary geometry and the inverse square law—that even in Olbers' day had been around for more than two centuries.

Mathematicians are, I think, more concerned than other scientists about the elegance of their work, and this may be why it seemed natural to start this book with mathematical or partly mathematical examples. But, though less attention may be paid to it, elegance can be found throughout the sciences, and I want to finish this chapter with two examples both almost entirely devoid of mathematics.

The first is an experiment published by William Harvey in 1628,[4] when he was about 50, though done about a decade earlier. At that time it was not realized that the heart was a pump, or that there was a continuous circulation of blood. Blood was thought to ebb and flow in an irregular manner, in both arteries and veins.

At the age of 19, having taken his BA at Caius College, Cambridge, and decided to become a physician, Harvey went to Padua, where he studied under Hieronymus Fabricius of Aquapendente. Fabricius had made a study of the valves whose presence in veins had been discovered in the previous century, but he misinterpreted the role of these valves, assuming that they were there merely to slow the flow of blood and so give the tissues more time to absorb nourishment. Having got his medical degree, Harvey returned to London, built up a medical practice and was appointed a physician at St Bartholomew's Hospital. In his spare time, he examined the heart and blood vessels of a vast number

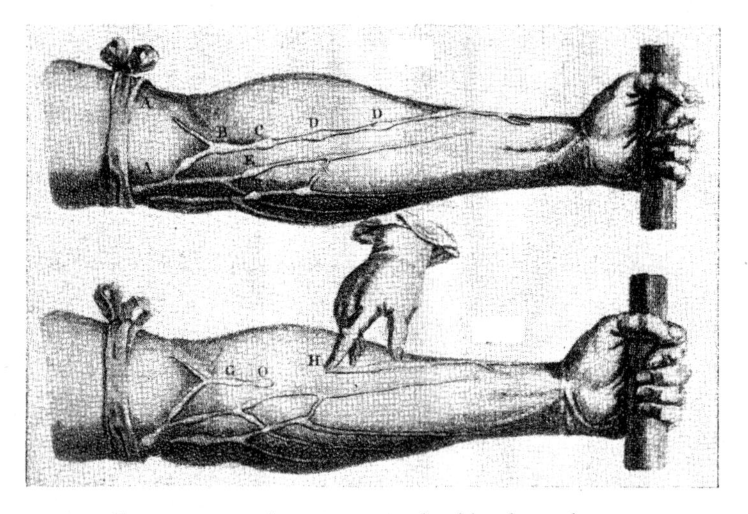

FIG 9 William Harvey's demonstration that blood circulates.

of animals, and did many experiments, but the experiment I want to describe did not require any animals and was extremely simple.

If you put a cuff around the human arm just above the elbow and make it tight but not too tight, the veins in the forearm become engorged, and you can see them as a network of slightly raised blue lines, with occasional 'knots or risings'—the phrase is Harvey's—indicating the presence of valves (see the upper part of Figure 9). If now you keep the tip of one finger firmly pressed on one point of a vein, and with a finger of your other hand you 'streak the blood upwards [i.e. towards the elbow] beyond the next valve,' you will find that the vein between the first finger and the valve remains empty even though the second finger has been removed (see the lower part of Figure 9). In Harvey's words, 'The blood cannot retrograde.' But as soon as you remove the first finger the empty portion of vein is immediately refilled from below. And this sequence of events can be repeated any number of times. The only possible conclusion from observations of this kind is that the blood in veins flows in only one direction, and that that direction is towards the heart. From the dimensions of the vein and the speed at which it was refilled, Harvey made a rough estimate of the rate of flow, concluding that not only does blood circulate, but that it circulates rapidly.

A remarkable feature of this experiment of Harvey's is that it is so easy to do that one has to ask why it was not done until the seventeenth century. Harvey himself was puzzled that Galen, who in the second century showed that, contrary to then current beliefs, the arteries contained blood not air,* had not gone on to discover the circulation. The long delay before the publication of Harvey's experiments no doubt partly reflects Harvey's caution in contradicting Galen, but there was also a real difficulty. If blood leaves the heart by the arteries and returns by the veins, how does it get from the arteries to the veins? It was not until after Harvey's death and the invention of microscopy that Marcello Malpighi discovered capillaries.

The diarist John Aubrey, writing about Harvey in his *Brief Lives*, says that Harvey told him that

> after his booke of the Circulation of the Blood came-out, that he fell mightily in his practize, and that 'twas beleeved by the vulgar that he was crack-brained; and all the physitians were against his opinion, and envyed him...

* The surprising early belief that arteries contained air arose from observations on dead animals. Erasistratus had noticed blood spurting from cut arteries in wounded animals, but had supposed that this was because the escape of *pneuma* [Greek for breath] from the damaged artery caused a vacuum, which sucked blood from the veins.

Be that as it may, when Harvey died in 1657 he was able to leave £20,000 to his brother.

The conclusion from Harvey's experiment was startling to his contemporaries but is not at all startling to us four centuries later. To end this chapter, here is a description of an observation that is still startling (though it was first made in 1898[5]), that reveals an important feature of our own vision and that, like Harvey's observation on veins, is something that you can confirm for yourself with almost no trouble. All you need is a mirror and a friend.

Ask your friend to stand about 60 cm in front of you and to look alternately at your left and your right eye, switching every second or two. You will find no difficulty in detecting the rapid flick of your friend's eyes at each switch. Now stand about 30 cm in front of the mirror (so that your reflection is about the same distance from you as your friend was) and look alternately at the left and the right eye of your reflection. You will see no movement at all—though an onlooker would have no difficulty in seeing the flicking of the eyes of your reflection. It seems that *while we are making rapid eye movements we ignore the images on our retinas.* This makes good sense since during those rapid movements the images on our retinas would be blurred and unhelpful, and would only complicate the analysis of the information we are receiving from the scene we are looking at. But to be able to 'switch off', as it were, is a remarkable ability, and one that is still not fully understood. It may involve some sort of 'gating' in the visual pathway, which prevents at least some incoming visual information from reaching the cerebral cortex from an eye that is rapidly changing direction.

In the rest of this book I want to look at examples of elegant theories, elegant explanations, and elegant experiments, in many different fields and coming from work done at very different times. I shall try to choose examples that are intelligible to the non-scientific reader, and to make clear what the 'state of play' was at the time the work being discussed was done. And because scientists are people, and often fascinating people, I shall sometimes try to give an idea of the history and character of the scientists themselves.

2

CELESTIAL MECHANICS: THE ROUTE TO NEWTON

> Nature and Nature's laws lay hid in night:
> God said, *Let Newton be!* And all was light.
> Alexander Pope, 1688–1744

Newton's laws of motion and of gravitation not only provided an elegant explanation of a vast body of astronomical observations that had been accumulating since the days of ancient Babylon, but also proved to be fundamental steps in the development of modern physics.

We are all familiar with some version of the story of Newton's apple, but to appreciate what Newton did we need to know the intellectual situation he faced. And to do that we need to look at the extraordinary way in which ideas about the Universe, which had remained almost static from the time of Ptolemy (second century AD) to the early sixteenth century, were developed by four remarkable, fascinating, and very different men over the next century and a half. The four men were Nicolaus Copernicus (a Pole), Johannes Kepler (a German), Tycho Brahe (a Dane), and Galileo Galilei (an Italian), and it is their work that provided the foundation for Newton's achievements in this area, and that also, incidentally, provides interesting examples of elegant and not-so-elegant science.

But we must start with Ptolemy—astronomer, mathematician, and geographer—for it was the massive treatise he produced in Alexandria that summarized the view of the Cosmos that came to be generally accepted for about fourteen centuries. Following the classical Greek tradition, he believed the Earth to be a perfect sphere at rest at the centre of the Universe, with the Sun, Moon, and planets moving round the Earth in concentric circular orbits that increased in diameter in the order: the Moon, Mercury, Venus, the Sun, Mars, Jupiter, and Saturn. The 'fixed stars'—so called because their positions *relative to one another* do not change—were thought of as being attached to the inside of an outermost sphere that was itself rotating with the Earth as its centre.

Although both Plato and Aristotle had been convinced that all heavenly bodies move in perfect circles at constant speed, this was difficult to reconcile with the observed behaviour of the planets, which sometimes seemed to speed up or slow down or even stop and reverse their motions. To explain this behaviour, later Greek astronomers, and Ptolemy himself, introduced several ingenious fudges that we need not consider here (but see note[6]).

Interestingly, Ptolemy regarded the fantastic system he ended up with—which contained 39 (theoretical) wheels; or 40 if you include the outermost sphere carrying the fixed stars—solely as a device for describing and predicting planetary movements. And for this they were rather successful, being (with only slight corrections) precise enough and accurate enough to be used for navigation by Columbus and Vasco da Gama in the fifteenth century. He did not think of them as a possible guide to the cause of these movements. Heavenly bodies, being of a divine nature, could not be expected to obey the same laws as bodies on earth.

The Ptolemaic system, accepted for so long, began to be doubted by astronomers in the questioning atmosphere of the early sixteenth century. The possibility that an apparent rotation of the fixed stars could be the result of a rotation of the Earth on its axis had been considered by Ptolemy but rejected as ridiculous. It was the work of Nicolaus Copernicus[7]—the Latinized form of Koppernigk (his father was a copper trader)—to reconsider that possibility and to revive another Greek theory, in which the Sun is the centre of the Universe.* Apart from the basic heliocentric idea there is nothing strikingly elegant about Copernicus' work—no elegant new theories or experiments or proofs. The agreement between predictions of planetary movements and observations of those movements still depended on two of the three kinds of fudge that Ptolemy had used, and initially the errors of predictions made using the heliocentric theory were greater. And there was a deeper worry. Copernicus had no doubt that the Earth moves, but the idea that the complicated theoretical machinery that was able to describe the movements of the Earth and all the planets corresponded to real machinery in real space was hardly tenable. And if it wasn't tenable his situation was uncomfortably close to that of Ptolemy in being able to describe and predict the movements of the planets without having any idea of what causes them to move or why they move in the way they do. He did not share Ptolemy's comforting view that heavenly bodies cannot be expected to obey the same laws as bodies on Earth. Yet, by reviving the heliocentric view, his great book, *De Revolutionibus Orbium Coelestium (On the Revolutions of the Heavenly Spheres)*, published in 1543 as he lay dying, was to prove enormously important.

* Proposed about 270 BC by Aristarchos of Samos. There are also Vedic Sanskrit texts from the ninth–eighth century BC in which the Sun is central and the Earth rotates round it. In the 5th century AD, work on the movement of the inner planets by the Indian, Aryabhata, may reflect a heliocentric model.

The solution to many of Copernicus' difficulties, and a much clearer understanding of the orbits of planets, would eventually be provided by Johannes Kepler.[8] In 1571, almost a century after the birth of Copernicus, Kepler was born in the little south-west German town of Weil der Stadt. His grandfather was mayor of the town; his father, who later deserted his family, would narrowly escape being hanged; his aunt, who brought up his mother, was burnt as a witch; and in her old age his mother, too, was accused of witchcraft and threatened with burning. Of his six younger brothers and sisters, three died in childhood and another was an epileptic whose life was a catalogue of disasters. He himself had been born prematurely, was often sickly as a child, nearly died of smallpox when he was 4, and had defective vision—possibly a result of the smallpox. But as an adult he remembered that when he was 6 he had been taken by his mother to see the 'Great Comet' of 1577, and when he was 9 he had seen an eclipse of the Moon and noticed that the Moon appeared quite red.

As a young child, Kepler attended school irregularly, but he profited later from the efficient school system (complete with scholarships for the male children of the 'poor and faithful') set up by the protestant Dukes of Württemberg with money confiscated from the monasteries; and by the time he was 20 he had graduated from the University of Tübingen. His interests had been philosophy, mathematics, and astronomy, but he now started a three-year course in theology, aiming at a career as a Lutheran minister. He was saved from that fate when partway through the course he was offered, and after some hesitation accepted, a teaching post in mathematics and astronomy at the Protestant School in Graz.

Although Kepler was a keen supporter of the Copernican system, his early work in Graz has a strong flavour of the

Pythagorean mystics who sought harmony in the celestial spheres. Drawing diagrams for his class one day, he suddenly realized that if you drew an equilateral triangle that just fitted inside a circle, and you then drew a smaller circle that just fitted inside the triangle (see Figure 10), the relative sizes of the two circles looked very much like the ratio of sizes of the orbits of Saturn and Jupiter. Continuing to muse, he wondered: if he drew a square inside the inner circle and a smaller circle just fitting inside the square, would the size of that smaller circle be proportional to the orbit of Mars? And could he repeat the process using a pentagon, a hexagon, and a heptagon to get the orbits of the Earth, Venus, and Mercury? He tried the experiment and the scheme didn't work. It then occurred to him that since the Universe was three-dimensional it was silly to use polygons: 'Why look for two-dimensional forms to fit orbits in space? One has to look for three-dimensional forms.'[9] And the obvious forms to use—because, as the ancient Greeks had shown, they are the only perfectly symmetrical three-dimensional solids (each one made with the same polygonal face repeated)—were the so-called Platonic solids (see Figure 11). There are five such solids: the tetrahedron (a three-sided pyramid bounded by four equilateral triangles); the cube (bounded by six squares); the octahedron (bounded by eight equilateral triangles); the dodecahedron (bounded by twelve pentagons); and the icosahedron (bounded by twenty equilateral triangles). The fact that there are only five Platonic bodies was particularly intriguing because there were just five spaces between the six planetary orbits,** so if each solid was to be used only once, and if Kepler's hypothesis was right, it would explain why there were only six planets (including the Earth).

** The Earth is included in the six. The planets beyond Saturn had not yet been discovered.

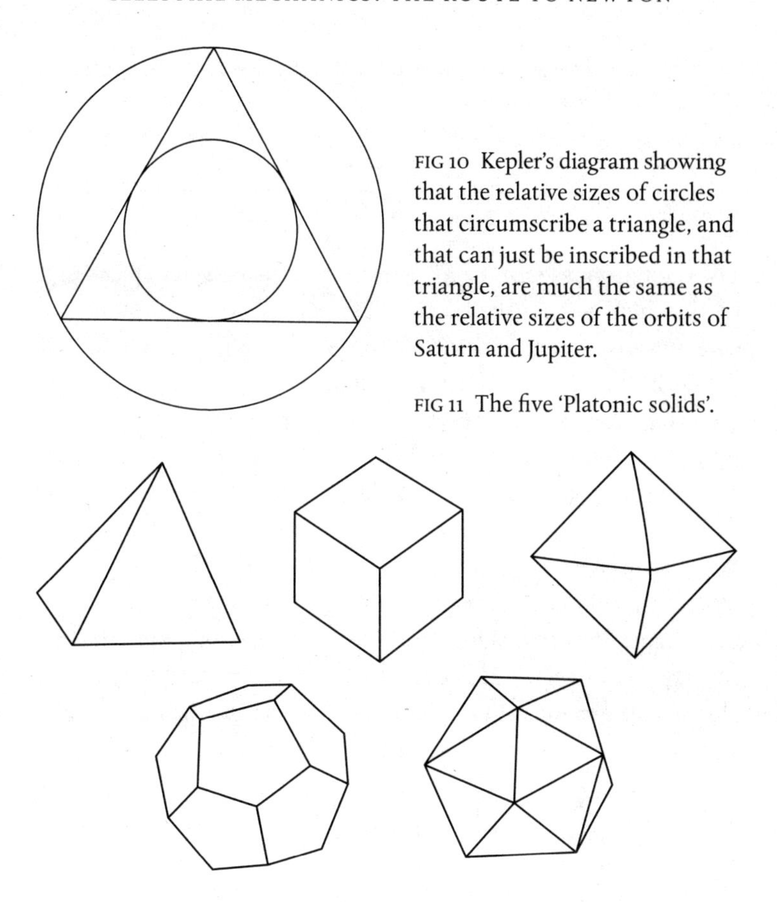

FIG 10 Kepler's diagram showing that the relative sizes of circles that circumscribe a triangle, and that can just be inscribed in that triangle, are much the same as the relative sizes of the orbits of Saturn and Jupiter.

FIG 11 The five 'Platonic solids'.

And when Kepler looked to see whether he could fit the five bodies neatly into the interorbital spaces he found that if he put the cube between Saturn and Jupiter, the tetrahedron between Jupiter and Mars, the dodecahedron between Mars and the Earth, the icosahedron between the Earth and Venus, and the octahedron between Venus and Mercury, the fit was not too bad.

> Day and night I spent with calculations to see whether the proposition that I had formulated tallied with the Copernican

> orbits or whether my joy would be carried away with the winds...Within a few days everything fell into its place. I saw one symmetrical solid after the other fit in so precisely between the appropriate orbits, that if a peasant were to ask you on what kind of hook the heavens are fastened so that they don't fall down, it will be easy for thee to answer him.[10]

The quotation is from the *Mysterium Cosmographicum*, Kepler's first book, which he published in 1597 at the age of 25, and it reflects his early enthusiasm about his conclusions. But in the second half of the book, he recognizes that the tallying with Copernican orbits is less good than he had at first thought, particularly for Mercury and Jupiter. His elegant (if somewhat mystical) solution to the inter-orbital space problem was already beginning to look doubtful—an early reminder that even the most elegant solutions can be wrong. What he needed was more reliable data about the planetary orbits, but that would have to wait for Tycho Brahe. Meanwhile he had raised questions that were both interesting and of a kind that no one had asked before. He pointed out that from existing data it is clear that the more distant a planet is from the Sun the more slowly it moves along its orbit, and he suggested that the Sun

> drives the planet the more vigorously the closer the planet is, but [the Sun's] force is quasi-exhausted when acting on the outer planets because of the long distance and the weakening of the force which it entails.[11]

By the end of the century Kepler was forced by the anti-Lutheran policy of Archduke Ferdinand to leave Graz. The first day of January 1600 found him on the road to Prague, accepting a long-standing invitation from Tycho Brahe, who had recently been appointed *Imperial Mathematicus* by Rudolph II, King of Bohemia and Hungary, and Holy Roman Emperor.

Twenty-four years older than Kepler, Tycho—he is usually referred to by his first name—was a Danish aristocrat, the son of the Governor of Helsingborg Castle,[†] but brought up by his paternal uncle who was a Vice Admiral.[12] At the age of 14 he observed a partial eclipse of the Sun and seems to have been amazed, not so much by the event itself as by the fact that it had been predicted; he immediately began buying books on astronomy. Two years later his uncle, anxious to counteract such distractions and get his nephew an education appropriate for a nobleman, sent him with a suitable tutor to Leipzig to study law; but though the young Tycho studied law during the day he continued to make astronomical observations secretly at night. The following year (1563) he observed the conjunction (i.e., apparent proximity[‡]) of Saturn and Jupiter, and noted that it was a month later than predicted by the thirteenth century *Alphonsine Tables*, and a few days later than predicted by the recent *Prutenic tables*.[13] He himself later said that this had been the turning point in his career. After three years at Leipzig he returned to Copenhagen, and shortly afterwards his uncle died, leaving Tycho free to devote himself to astronomy.

The inaccuracy of the predictions of the conjunction of Saturn and Jupiter had made him realize the need for much more accurate observations, and he greatly improved the instruments that astronomers then used for naked-eye observing. While working in Augsburg he had a quadrant—a device for measuring angles up to 90°—made with a radius of about 19 feet.

On the evening of 11 November 1572, he was walking home to supper when he noticed an extra star—and one brighter than all the others—in the constellation Cassiopeia, then almost directly

[†] Helsingborg is the town across the Sound from Elsinore.

[‡] They were not really close, but because they and the Earth were almost in line they looked close.

overhead. To measure the star's angular distance from the other stars in the constellation, he used a huge sextant made of seasoned walnut with a bronze hinge, taking elaborate precautions to ensure that the sextant remained vertical and stationary during each set of measurements. He observed the star over the next sixteen months, as it gradually became fainter, and found that there was no detectable change in the angular distances between the new star and the nine normal members of the constellation. At that time it was firmly believed (following Aristotle) that changes occur only in the sublunar region of the heavens, but Tycho calculated that if the new star was no more distant than the Moon there would have been an annual stellar parallax of nearly one degree—easily detectable with the apparatus he was using. (Parallax is the apparent shift of an object against a background due to a change in the position of the observer. The nearer the object is to the observer, the greater the parallax.) It followed that something pretty dramatic was happening in a region of the heavens that was supposed to be unchanging. What he was seeing—what he called the *nova stella*—is what we should now call a *supernova*; they occur rarely, on average not more than once or twice a century in our galaxy, and are the result of thermonuclear explosions releasing quite prodigious amounts of energy.

A side-effect of the 1572 supernova was that it made Tycho famous, and a few years later Frederick II of Denmark offered him the use, for the rest of his life, of the island of Hveen—an island of more than 2,000 acres, a few miles south of Elsinore—together with the 'rent and duty' from all the Crown's tenants and servants on the island, a generous annual grant, and the cost of building a house and observatory.[14] The offer was irresistible, and Tycho built an observatory which Koestler, judging from surviving woodcuts, describes as looking 'rather like a cross between the Palazzo Vecchio [in Florence] and the Kremlin.' Here Tycho lived, and worked very successfully as an astronomer, for the next twenty years, though he could never quite swallow the Copernican system—planets might revolve round the Sun but the Earth had to be at the centre of things. Unfortunately, Tycho was an arrogant and despotic landlord, and when Frederick II was succeeded by Christian IV, relations gradually deteriorated, and in 1597 he felt he had to leave Denmark. By the turn of the century, the Emperor Rudolph had appointed him *Imperial Mathematicus*, with a salary—in theory at least—of 3,000 florins a year, and with a castle to play with at Benatky, twenty-two miles from Prague. This was where he met Kepler.

The conjunction of Kepler and Tycho in 1600 was as significant for Kepler's career as the conjunction of Saturn and Jupiter in 1563 had been for Tycho's. At the age of 28, Kepler had already produced his remarkable theory to account for the spacing of the planetary orbits—a theory that didn't quite work—and he was asking novel questions, and making novel suggestions, about the causes of the movements of heavenly bodies—questions and suggestions of a kind that would in time lead to major advances, but hadn't yet. What he needed, and what he knew he needed, was more accurate data about the planetary orbits, and no one was better placed than Tycho to provide them.

Kepler met Tycho in Benatky in early February, but though each was aware how useful the other could be there were complications: Kepler needed an income and somewhere to live; Tycho was having trouble getting his promised salary, and was still without some of his assistants and instruments. In the ensuing difficult discussions with Tycho, Kepler's mood oscillated between anger and deep contrition, but after a reconciliation he settled down to work. At Tycho's request he took over the study of Mars from one of Tycho's assistants, who was not making much progress. A month later he returned to Graz, primarily to collect his wife, but while there he continued to work on the problems of determining the orbit of Mars, and he developed a device for safely observing a solar eclipse, which he used in the Graz market square. By the time he returned to Bohemia, Tycho had moved his centre of operations from Benatky to Prague, to satisfy Rudolph's desire to have his *Imperial Mathematicus* closer at hand—presumably to give astrological advice. For the next few months Kepler continued to work on Mars, but a letter written to his former teacher at Tübingen in February 1601 reveals his irritation:

> Tycho is very stingy as to communicating his [astronomical] observations...I am allowed to use them daily. If I could only copy them quickly enough! I must, however, be content with making selections from them...All [his observations] are accessible to me but first I had to promise solemnly to keep them secret.[15]

Towards the end of October the situation changed dramatically, when Tycho died from acute urinary retention. Two days after the funeral, the Emperor appointed the 29-year-old Kepler *Imperial Mathematicus*, a post he retained to the end of his life. Kepler seems to have reacted to Tycho's death as promptly as the Emperor. 'I confess', he says in a letter to a correspondent in England, 'that when Tycho died, I quickly took advantage of the absence, or lack of circumspection, of the heirs, by taking the observations under my care, or perhaps usurping them...'[16] The heirs were naturally furious and there was trouble for some years, but Kepler had got what he needed.

With the detailed and accurate information available in Tycho's data, Kepler had expected to be able to describe the orbit of Mars in about a week. In fact it took him the best part of five years, though when he had finished he had discovered more than just the orbit. Part of the delay was caused by two distractions.[17] First, he spent a great deal of time investigating optics and writing a 450-page book on the subject.[18] The second distraction was the appearance in 1604 of a new star (another *supernova*), which was first seen near the conjunction of Saturn and Jupiter, with Mars not far away. As *Imperial Mathematicus* he was expected to explain how the new star originated—an astronomical problem—and what it signified—an astrological problem. His approach to the two problems was very different—wildly speculative for the astrology; generally hard-headed for the astronomy. Because Kepler's new star did not change its position relative to other fixed stars, he rejected the suggestion that it had been ignited by the planets, and assumed that it was the result of an agglomeration of heavenly material resulting from an 'architectonic natural ability' inherent in that material. When it was suggested that the agglomeration might merely have been the result of a fortuitous accumulation of atoms he replied:

> I will tell these disputants, my opponents, not my own opinion but my wife's. Yesterday, when weary with writing, and my mind quite dusty with considering these atoms, I was called to supper, and a salad I had asked for was set before me. 'It seems, then,' said I aloud, 'that if pewter dishes, leaves of lettuce, grains of salt, drops of water, vinegar, and oil, and slices of egg, had been flying about in the air from all eternity, it might at last happen by chance that there would come a salad. 'Yes,' says my wife, 'but not so nice and well dressed as this of mine is.'[19]

Despite these distractions, by 1606 Kepler had succeeded in solving the problem of the orbit of Mars. He had started using the sort of technique that Ptolemy and Copernicus had used, but after an immense amount of work he concluded that the orbit could not be circular and that the speed of Mars round its orbit could not be constant. But if the orbit was not circular what was it? This was a major difficulty because to answer the question he needed to be able to discover the position of Mars at each point on its orbit, but all observations of Mars had been made from the Earth, which was itself moving. To determine the actual movement of Mars then, he first needed to know precisely how the Earth was moving. To discover the position of the Earth at any moment, he needed to do what a navigator does when his ship is some way from the coast but he is not sure where: he takes the bearings of two fixed points, draws lines on the map from those points, and reckons that the ship must be at the intersection of the two lines. The trouble was that there was only one obvious fixed point—the Sun. (The fixed stars could not be used because they were so far away that their bearings did not change perceptibly as the Earth went round its orbit. They did, though, serve to orient the whole solar system in space.)

Faced with this problem, Kepler had a solution that was so elegant and so counter-intuitive that Einstein, over three centuries later, called it 'an idea of true genius'.[20] He would use Mars as his second fixed point. Though Kepler could not predict the exact position of Mars at any time, he did know that it takes 687.1 Earth days for Mars to go round its orbit. So every 687.1 Earth days (one Mars year) Mars would be in exactly the same position. By selecting, from the mass of Tycho's data, sets of observations taken at intervals of exactly one or more Mars years, and restricting his attention to those data, he had his two fixed points and could calculate the position of the Earth at those times. Having determined the precise shape of the Earth's orbit and the way the Earth moved along it, he could use observations of Mars to determine Mars' orbit and the way Mars moved.

The upshot of this work and a great deal more was the recognition by Kepler that three rather simple laws were sufficient to describe planetary movements.

Kepler's **first law**, about which he was initially uneasy because it involves an asymmetrical arrangement, is that each planet moves in an elliptic orbit[§] with the Sun at one of the foci—see Figure 12. How much the elliptic orbit departs from a circle differs for different planets. The orbit of the Earth is not far from a circle; the orbit of Mars is the least circular of all the planetary orbits—which is why Kepler and his predecessors had such difficulty in trying to predict Mars' movements on the assumption that it moved round a circular path.

His **second law** is that the speed at which a planet moves round its orbit is not constant but increases as the planet approaches the Sun and decreases as it recedes from it. More precisely, as a planet moves along its orbit, a line drawn from the planet to the Sun sweeps out equal areas in equal times—see Figure 13.

His **third law** (not discovered till 1618, when he was 46) is that if you compare different planets you will find that the square of the time it takes each planet to complete one orbit is proportional to the cube of the length of the major axis of the ellipse (see Figure 13). Since his youth, Kepler had wondered what the relation was between a planet's distance from the Sun and the time it took to complete its orbit round the Sun; now he had an answer.

[§] An ellipse is an oval-shaped curve such that the sum of the distances from any point on the curve to two fixed points is constant. The two fixed points are called foci (plural of focus). If you knock two nails some distance apart into a horizontal wooden board, place around the nails a loop of string of length well over double the distance between the nails, hold the string taut with a vertically held pencil, and, continuing to hold the string taut, move the pencil through 360° around the pair of nails, you will have drawn an ellipse.

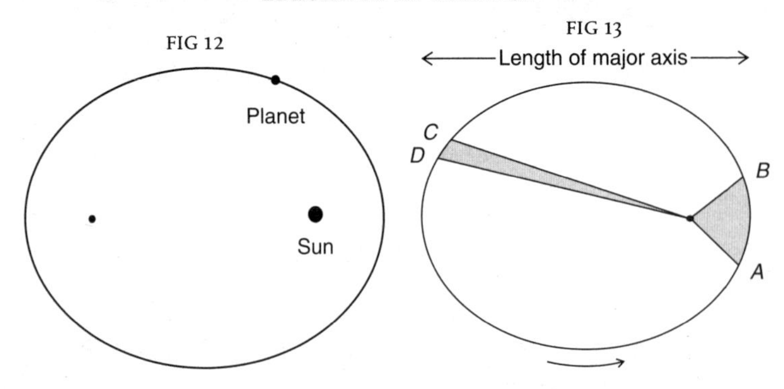

FIG 12 An elliptic orbit with the Sun at one of the foci.

FIG 13 Diagram illustrating Kepler's second law.

Using Tycho's data, Kepler had not only succeeded in discovering just how planets move but he had been able to describe their movements using three simple mathematical formulae. This was elegant economy indeed. The next step—understanding why planets move in this way—would have to wait for Newton, but Kepler did make some suggestions. Though not correct, these suggestions were impressive because they were based on known physics rather than on fantasies about platonic bodies (as in his early work) or on analogies between the mathematical patterns that the Pythagoreans had shown were associated with musical harmony, and the mathematical patterns that might be associated with the 'harmony of the Universe' (as in his later work). Early on, he had thought it likely that the Sun provided the force that made the planets rotate, but to account for their rotation he supposed that it somehow acted like a sweeping broom; indeed he wondered whether the Sun itself rotated on its own axis—and some years later Galileo's observations of sunspots showed that it did.

He was aware of the idea of gravitational attraction between bodies. Not only did he suggest that the Moon's gravity was the cause of tides, but in the preface to his *Astronomia Nova*[21]—the book, published when he was 38 that contained his first two laws of planetary motion—he wrote:

> Gravity is the mutual bodily tendency between cognate[#] bodies towards unity or contact…, so that the Earth draws a stone much more than the stone draws the Earth…
>
> If two stones were placed anywhere in space near to each other, and outside the reach of force of a third cognate body, then they would come together, after the manner of magnetic bodies, at an intermediate point, each approaching the other in proportion to the other's mass.[22]

His difficulty was that, in the absence of the modern concepts of *momentum* and *inertia*—though he would later introduce the word *inertia* with a more restricted meaning—he could not see how a gravitational attraction alone—or, for that matter, a magnetic attraction to the Sun—would make the planet rotate round the Sun. (This was just nine years after William Gilbert had shown that the Earth itself acts like a giant magnet.)

Important as Kepler's first two laws would prove to be, they caused little stir at the time. For the next two years, Kepler turned his attention to improving telescopes—which had been invented by spectacle-makers in Holland in 1608 and had enabled Galileo to discover the moons of Jupiter. In 1611, Kepler lost his wife from 'Hungarian fever' (probably typhus) and his 6-year-old son from smallpox, and in the following year he had to flee Prague: Lutherans had now become unacceptable there. The next fourteen years he spent at Linz—marrying again and having further children; writing a long and successful account of Copernican astronomy; defending his mother during her prolonged trial for witchcraft; and (reverting to his more

[#] From the Latin *cognata*, akin in origin, nature, or quality.

mystical mode) writing his book *De Harmonices Mundi* (*Harmonies of the World*)—which did, however, contain his totally unmystical third law in the last chapter. In 1626, caught up in the Thirty Years' War, Kepler made his third move on religious grounds.## He moved to Ulm, in Württemberg, where at last, twenty-six years after Tycho's death, he published the long awaited and much acclaimed *Rudolphine Tables*—the first modern astronomical tables, which Tycho had promised to prepare for the Emperor, and which Kepler had promised the dying Tycho that he would complete.

Before leaving Kepler though, I ought to mention something else. Although he lacked the modern concept of *inertia*—the tendency of a body to retain its current velocity unless acted on by an unbalanced force—in his book about Copernican astronomy he did introduce the word *inertia* (from the Latin word for idleness or laziness) in a more restricted sense. He used it as a measure of a resting body's tendency to remain at rest. A modern use of the word in this restricted sense would be to say that if you put a ping-pong ball and a billiard ball on a smooth horizontal surface, and flick both as hard as you can, the ping-pong ball will shoot across the surface, while the billiard ball will move sluggishly because it has a much greater inertia. It was only when Galileo (though not actually using the word inertia) extended the meaning of the concept to refer to a measure of a body's tendency *to retain the speed and direction of movement that it currently has*, that the concept began to make a major contribution to our understanding of movement, and therefore of planetary movement.

Linz, the town he left, would three centuries later produce Hitler. Ulm, the town he moved to, may have been where Descartes had his famous dreams in 1619, and was certainly the town where Einstein was born in 1879.

Galileo's contribution can wait until the next chapter, but I want to finish this one by considering what it was that led Kepler to make the statement (quoted above) about the effect of the mutual attraction between two stones, near each other but isolated in space. That sentence, which comes rather close to a law of universal gravitation, states that the two stones:

> would come together...at an intermediate point, each approaching the other in proportion to the other's mass.

This behaviour, we now know, would follow from Newton's second and third laws of motion, but Kepler wrote the sentence more than thirty years before Newton was born. It is conceivable that for some reason he felt (correctly) that, isolated from other influences, the stones would move in such a way that the centre of mass¶ of the isolated system would not change; but that too would show a remarkable insight, and if it had been the basis of his statement you would expect him to have mentioned it. A more likely hypothesis, suggested to me by my colleague Piero Migliorato, is that he was looking for an explanation of the asymmetry of the situation of a falling stone. Two stones, near each other but isolated in space, approach each other; but when a stone falls towards the Earth, the Earth shows no detectable rise towards the stone. It seems natural to relate that asymmetry to the enormous asymmetry between the mass of the stone and the mass of the Earth. And there is no doubt that Kepler did believe that the Earth rises slightly towards the stone. In a letter, written in 1605 to a colleague in Lower Saxony, he says:

¶ The concept of a centre of mass (or a centre of gravity) goes back to Archimedes.

> ...not only does a stone approach the Earth, but the Earth also approaches the stone, and they divide the space between them in the inverse ratio of their weights.'[23]

If a body's inertia—in the limited sense used by Kepler—is proportional to its mass, the difference in the inertia of the falling stone and of the Earth provides a simple explanation of the asymmetry of the situation of the falling stone. Similarly, if inertia is proportional to mass, two stones near each other and isolated in space would be expected to move towards each other, each moving a distance proportional to the other's mass. Although Kepler didn't write about inertia until 1618, it looks as though he must have been thinking about it more than a decade earlier.

3

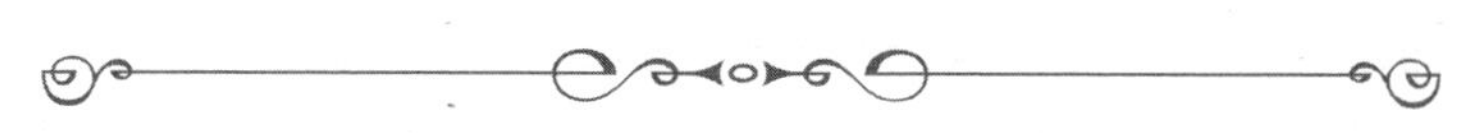

BRINGING THE HEAVENS DOWN TO EARTH

It was Galileo's novel and elegant approach in investigating the movements of terrestrial bodies that eventually made it possible to link terrestrial and celestial movements.

Galileo Galilei was born in Pisa in 1564 to a poor branch of a patrician Florentine family.[24] His father, Vincenzo, was a composer, a fine lute player, and a writer on the theory of music who showed, for example, that the musical interval of a perfect fifth could be produced not only by having two strings under the same tension with lengths in the ratio 3:2, but also by having two strings of the same length supporting weights in the ratio 4:9 (i.e., the square of the inverse ratio). The taste for a combination of mathematical analysis and direct experiment seems to have been inherited by his son. Vincenzo would have liked his son to become a businessman and retrieve the family fortune, but Galileo's obvious intellect made a university course desirable; so, knowing the poor financial rewards of mathematics and music, his father entered him for the medical course at Pisa. In his first year, while visiting Pisa cathedral, his mind ranging 'far and fast while riding to the anchor of a liturgy',* Galileo noticed that an oil lamp that had

* The quotation is from a sermon preached in 1932 by the Revd F.A. Simpson, a Fellow of Trinity College, Cambridge, and at one time often talked of (fortunately wrongly) as the last Cambridge eccentric.

been set swinging kept the same frequency as the swings became weaker and weaker; later he would make a study of pendulums and show that the frequency of a pendulum is independent of the weight of the bob. At this time he knew very little mathematics but a chance attendance at a geometry lecture given by a friend of his father inspired him, and a shortage of funds and his failure to get a scholarship to study medicine led to his leaving Pisa at the age of 21 without a degree. He returned to Florence, where his family was now living, and studied mathematics privately; four years later he was appointed a lecturer in mathematics by the university that had refused him a scholarship. Back in Pisa he began the study of motion—particularly terrestrial motion—which, with the associated demolition of the still prevailing Aristotelian physics, was to be his most important contribution to science.

Aristotle had distinguished between 'natural motions', such as a heavy body falling to earth, or air drifting upwards, and 'violent motions', such as an arrow shot from a bow, or a horse pulling a chariot. All motions, he believed, whether natural or violent are caused by 'motive forces', forces which must, he thought, be acting as long as the motion continues. And stronger forces produce faster motions. Two horses pull a chariot faster than one, and a heavier body, he assumed (not always correctly), falls more quickly than a lighter one. Indeed, he went further and declared (wrongly) that the rate at which a body falls is proportional to its weight.

He had recognized one difficulty; if the movement continues only so long as the motive force acts, why does the arrow continue to fly when it has left the bow? He had to assume that the surrounding medium plays a part; the bowstring pushes the medium as well as the arrow and, for a while, the medium must take over from the bowstring. The medium involved could not be air, he thought, because air could not cause a heavy object to move adequately; some more mysterious medium had to be involved.

By the time of Galileo, these Aristotelian notions had been around for nearly 2,000 years, and some of them seemed just common sense. But the statement that the rate at which bodies fall is proportional to their weight could readily be proved wrong by a straightforward 'real experiment' and could be shown to lead to an inconsistency by an elegant 'thought experiment'. Galileo was responsible for both.

In the real experiment, he dropped a cannon ball weighing more than a hundred pounds and a musket ball weighing half a pound from a great height, and they reached the ground virtually at the same time. It has long been believed that this experiment was done—possibly as a theatrical public demonstration—from the Leaning Tower of Pisa, and, given that Galileo was working in Pisa when he started this sort of work, the belief is plausible, though there have been suggestions that the story has been 'improved'.

The thought experiment was not theatrical but in its simplicity, ingenuity, conciseness, unexpectedness, and persuasiveness it was highly elegant.[25] Suppose two bodies of very different weights are tied together by a length of string, and then dropped from a height. If Aristotle is right, the heavy weight will fall faster and will tend to increase the speed of the lighter, and the lighter will fall more slowly and will tend to reduce the speed of the heavier; the speed at which the joined pair falls will therefore be intermediate between the speeds at which the two bodies would fall if they were separate. But if Aristotle is right you can also argue that the weight of the joined pair is greater than the weight of either alone, so the joined pair should fall faster than either body would fall alone. There must be something wrong with a set of assumptions that leads to incompatible conclusions.

It is easy to understand how Aristotle could have assumed that heavy bodies fall more quickly than light ones—a dropped coin falls much more quickly than a dropped feather—but it is extraordinary that his view about the proportionality of weight and speed of fall could have survived for nearly two millennia. Not everyone has a leaning tower handy, but it is easy to take two coins of very different weights, to drop them simultaneously, and to see them hit the floor at the same time.

Though Galileo's arguments were convincing, neither they nor his sarcastic manner endeared him to the traditional Aristotelians in the Philosophy Faculty; and when he irritated the Medici family by criticizing a scheme for dredging the harbour of Leghorn, and irritated the University by writing a jocular poem about academic dress, he began to find his position in the University neither comfortable nor secure. At the end of his three-year contract he left Pisa, and with the help of his friend, patron, and fellow mathematician, the Marchese Guidobaldo del Monte, he was appointed Professor of Mathematics at Padua. He was then 28, and would remain at Padua for eighteen highly productive years.

Padua was part of the Venetian Republic, and its University, already nearly 400 years old, was relatively free from domination by the Roman Catholic Church, and attracted students from all over Europe. (William Harvey visited it while Galileo was there.) Galileo settled down to teach geometry, mechanics, military engineering, and astronomy, and he continued with his work on motion. Having proved that Aristotle was wrong about freely falling bodies, he had to face the question: how fast do such bodies fall, and how does their speed change as they fall? The difficulty was that freely falling bodies move so fast that there was no obvious way of following their motions.

Instead of working with freely falling bodies, he therefore investigated the way smooth heavy bronze balls accelerated as they rolled down inclined polished wooden planes. By making the slope gentle—sometimes just under 2°—he could make the speed slow enough to measure. One of his most successful methods of measurement was to take advantage of human sensibility to any deviation from a regular rhythm. If you are listening to regular beats half a second apart, nearly everyone can detect an error if one beat is just $\frac{1}{32}$ of a second too early or too late. What Galileo (a lute player himself) did was to tie loops of catgut at intervals round the wooden slope—rather like the gut frets tied round the neck of a lute. To set the initial position of the loops he released the ball in time with the beat of a tune with a strong rhythm, and marked roughly the position of the ball at each subsequent beat. He then placed the loops so that there was a thread of gut crossing the path of the ball at those marked points. The sound of the rolling ball crossing the gut was just audible, so in subsequent runs he could adjust the position of the loops until the sounds of the crossings were precisely in time with the music. The distance between successive loops then told him the distance travelled by the ball in successive identical intervals of time.[26]

What he found astonished him. If the distance travelled in the first interval was defined as 1 unit, that travelled in the second interval turned out to be 3 units, in the third 5 units, in the fourth 7 units, and so on—a 'law of odd numbers'. When he compared the total distance travelled with the time the ball had been travelling, he found that the distance was proportional to the square of the time. If the behaviour of balls rolling down slopes was a guide to objects falling freely, then at last he had his 'law of free fall'; but what did it mean? It was only some years later that he realized that both the odd-number law and the proportionality between the distance travelled and the square of the time taken were simply the result of the fact that as the ball rolled down the slope

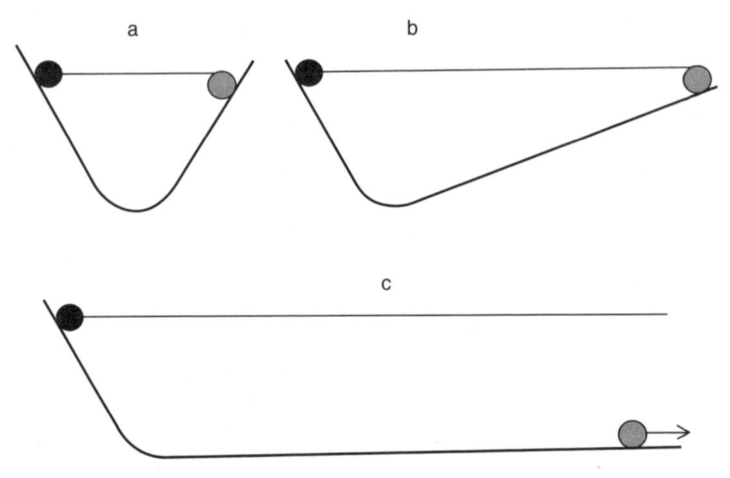

FIG 14 Galileo's experiments, with balls rolling down smooth almost friction-free planes, that led him to the concept of inertia.

its speed increased linearly with time. In other words, the acceleration was constant.

Elegant and important as these experiments were, they are rivalled by another set of experiments that Galileo did with balls and inclined planes, which led him to the concept of inertia—though he did not use that word. In experiments in which pairs of inclined planes faced each other, he found that a ball that had rolled down one plane and was then rolling up the other almost reached the height from which it had started; this was true even if the slopes were different—see Figure 14. And the more that friction could be reduced, the closer the ball got to its original height. If the second plane had a gentler slope than the first, the ball would have to travel further along it to approach its original height; and if the second slope were made successively more and more gentle, the ball would travel further and further. In fact, Galileo concluded, if the second plane were horizontal and if friction could be eliminated the ball would travel along at constant speed as far as the plane extended. This not only contradicted the idea that a force must be acting as long as a motion continues, but it also came close to anticipating Newton's first law of motion—sometimes known as the law of inertia—*that a body continues in its state of rest or of uniform motion in a straight line unless acted on by a force.*

Having established the 'law of free fall' and the concept of inertia, Galileo, with a slight modification, took these experiments with balls and inclined planes even further, to yield two more important conclusions. As before, he allowed balls to run down an inclined plane onto a horizontal plane, but the horizontal plane was now the surface of the table, and when the ball left the edge it curved downwards to the floor. By altering the height on the inclined plane at which the ball started he could vary the speed at which the ball rolled along the table surface. The interesting questions were: how did the horizontal speed on the table affect the path of the ball to the floor? And for any one speed what shape was the path to the floor?

What he expected was that on reaching the edge of the table the ball's steady horizontal motion would continue unchanged, but that the ball would also acquire the downward vertical motion associated with free fall, i.e. with the distance fallen proportional to the square of the time. If the ball's horizontal displacement from the edge of the table is proportional to the time that has elapsed since it left that edge, and the vertical displacement is proportional to the square of that time, the path of the ball to the floor must be half a parabola.[27] And that is precisely what he did find—see Figure 15. This was important in two ways. Firstly, it showed that the motions of an object in two directions perpendicular to one another can be changed independently of each other. Secondly, it introduced the then novel idea that the flight

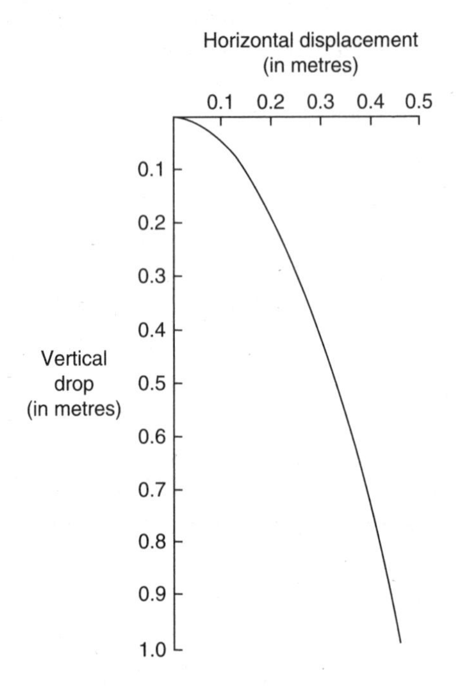

FIG 15 Galileo's experiment showing parabolic motion.

path of projectiles is parabolic and the result of the combination of motions induced by different forces. To reach so many important conclusions by experiments of such simplicity was not only very elegant science but it was science of a novel kind—making mathematical descriptions of real motions rather than trying to fit real motions into preconceived notions of perfect circles or whatever.

Although most of the work I have been discussing was done in Padua by 1609, little of it was published until near the end of Galileo's life. (This, incidentally, is why Kepler could publish a less sophisticated account of inertia in about 1620.) The cause of the delay was largely that Galileo (like Kepler) was distracted from his work on dynamics by the reports of the invention of the telescope by Dutch spectacle-makers. Visiting Venice in 1609, he heard that the reports were true and that a foreigner bringing one to Italy had recently arrived in Padua. By the time he got back to Padua the foreigner had left, so Galileo, knowing the basic design of the Dutch invention, decided to make a telescope himself. He not only succeeded but eventually achieved a thirty-fold magnification, which is about the best that can be achieved with the Dutch system, which used a convex object lens and a concave eyepiece.** He started using the telescope in early January 1610, and by the end of March he had published his *Sidereus nuncius* (*Starry Messenger*), a forty-eight-page octavo booklet describing mountains and craters on the Moon, revealing that the Milky Way was made up of myriads of stars, describing a host of new fixed stars, and, most remarkable of all, reporting that the planet Jupiter had four moons. The Earth, with its Moon, no longer seemed the odd one out in the heliocentric Copernican system.

An early objection to that system was that if the Earth were rotating daily, anyone on the surface (unless they were near the North or South Pole) would be moving very fast indeed—just

** This is the system that later was much used for opera glasses. It does, though, give a rather small field of view, which sometimes added to Galileo's difficulties when he was trying to demonstrate what was to be seen to untrained users. Nevertheless, Milton, who visited the aged Galileo in 1638 was impressed enough to write in *Paradise Lost*:

> ... like the Moon whose Orb
> Through Optic Glass the *Tuscan* Artist Views
> At Ev'ning from the top of *Fesole*,
> Or in *Valdarno*, to descry new Lands,
> Rivers or Mountains in her spotty Globe.

over a thousand miles per hour for someone on the Equator, and nearly three quarters as fast for someone in Padua. But if the Earth's surface were moving at such speeds, you might expect that anything not attached to it (such as an apple falling from a tree, or a stone thrown upwards into the air) would rapidly be left behind.

Before Galileo, the standard way of dealing with this kind of objection was to refer to the kind of thought experiment suggested by the English astronomer Thomas Digges, who in 1576 had pointed out that a stone dropped from the mast of a fast moving ship lands at the base of the mast, not further back. A more elaborate and more elegant variation of this thought experiment was described by the Italian, Giordano Bruno—Dominican monk, Copernican astronomer, expert on mnemonics—who was burnt at the stake in 1600. In one version of Bruno's thought experiment we are asked to think of a ship with a tall mast moving fast along a canal crossed by a bridge just high enough to clear the mast. If one person drops a stone from the top of the mast as it reaches the bridge and at the same moment another drops a stone from the bridge, the stone from the mast will fall on the deck immediately below the mast; the stone from the bridge will land further back on the deck. In other words, the stone that was moving forwards with the ship continues to move forwards during its fall; the stone that was not moving forwards with the ship does not move forwards during its fall. By analogy, the apple falling from a tree on a fast-rotating Earth will continue to move in the same direction during its fall and (if not interfered with) will end up precisely below the branch from which it fell.

Galileo's experiments demonstrating inertia provide a straightforward explanation of the behaviour described in these thought experiments, and so provide a sound basis for dismissing the traditional objection to Copernican theory; yet, probably through fear of ridicule, Galileo continued to teach the Ptolemaic system. Even his *Starry Messenger* did not explicitly recommend the Copernican view. Despite this reticence, the effect of the book was instant fame, and almost instant controversy. Although the Venetian senate offered to double Galileo's salary, he resigned his professorship at Padua and accepted the post of Mathematician and Philosopher to the Grand Duke of Tuscany, and Chief Mathematician of the University of Pisa with no teaching duties—a post which still sounds attractive 400 years on. Later in 1610, Galileo, using his telescope, saw the phases of Venus and realized that they were readily explicable on the Copernican system but completely inexplicable on the Ptolemaic. In 1613, in a book on sunspots, he talked for the first time about what would now be called rotational inertia—the ability of a rigid body *rotating about an axis passing through its centre of mass* to maintain its rotation indefinitely in the absence of friction or other forces—and he argued strongly (and for the first time in print) for the Copernican system. Curiously, at no time in his life did he show any interest in Kepler's elliptical orbits or his planetary laws; he appeared to be wedded to circular orbits.

Galileo's sudden enthusiasm for Copernicus seems to have been the result of a conscious decision, for he wrote to the Grand Duke's secretary that it was on the establishment of the Copernican doctrine that 'all his life and being henceforward depended.'[28] This turned out to be only too true. Late the following year, he was denounced from a pulpit, and the protracted tragedy of his later life began. In 1616 the Church decreed that the views of Copernicus were heretical, but in 1624 Galileo got permission to present the cases for both heliocentric and geocentric theories in an impartial way. Over the next six years he wrote his *Dialogue Concerning the Two Chief World Systems*. Written in Italian as a dialogue between three characters—one representing the author, one an intelligent listener, and the third well-meaning but called Simplicio—the book flagrantly ignored the conditions associated with the permission Galileo had received. In 1633 he was tried for heresy and sentenced to life imprisonment, subsequently commuted to house arrest. It was at his villa in Arcetri, near Florence, that he wrote *Discourses and Mathematical Discoveries Concerning Two New Sciences,* his last and major book, containing in its final section his crucially important work on motion, much of it done more than two decades earlier. The Congregation of the Index had banned the printing of any of his books, so the manuscript had to be smuggled out of the country and was eventually printed in Leiden in 1638.

Galileo died in January 1642; on Christmas Day that year, Newton was born.

By 1642, nearly a century had passed since the publication of Copernicus' *On the Revolution of the Heavenly Spheres,* and the intellectual world into which Newton was born was very different from that with which Copernicus had had to struggle. Kepler's three laws of planetary movement provided a satisfactory account of the ways planets behaved, and Galileo's laws of free fall and of projectile motion (with its concept of inertia) not only demolished the arguments against the notion that the Earth moved but also knocked Aristotle off his pedestal and helped to make physics an experimental science. Newton would produce an extraordinarily elegant explanation of all those five laws by showing how they depended on three very simple laws of motion and one very simple law of gravitational attraction; but he first had to survive a childhood as fraught as Kepler's.[29]

His 36-year-old father, who owned Woolsthorpe Manor—a modest stone farmhouse in Lincolnshire—and was a successful (though illiterate) yeoman farmer, died unexpectedly nearly three months before Isaac was born. The birth was premature, and Isaac himself was later told that when he was born

> he was so little they could put him into a quart pot & so weakly that he was forced to have a bolster all round his neck to keep it on his shoulders and so little likely to live that when two women were sent to...North Witham for something for him they sat down on a stile by the way and said there was no occasion for making haste for they were sure the child would be dead before they could get back.[30]

When he was three his mother married an elderly widower neighbour, the rector of North Witham, and moved to his rectory, leaving the young Isaac at Woolsthorpe to be looked after by his maternal grandparents, for whom he never expressed much affection. As a young child, he went to two local 'dame schools', and when he was 11 his stepfather died and his mother returned to Woolsthorpe with three children from her second marriage. A year later he became a pupil at the long-established King's School at Grantham—the town that was to be the birthplace of Margaret Thatcher—where he lived with a local apothecary in a household that included three of the apothecary's stepchildren. Isaac apparently got on well with the girl, for whom he made dolls' furniture, but not with the boys either at the apothecary's house or in school. He was keen on drawing—birds, animals, ships, plants, Charles I, his schoolmaster—and on making mechanical toys—a mouse-powered mill, fiery kites, clocks, a four-wheeled vehicle driven by a crank that he turned as he sat in it. And he was obsessed by sundials. As a pupil, he oscillated—often falling behind, then shooting ahead. Apart from biblical knowledge, he learnt a great deal of Latin, a little Greek, and probably very little mathematics.[31]

When he was nearly 17 his mother brought him back from Grantham to help manage the farm—a task for which he had neither interest nor aptitude, preferring to spend his time building gadgets rather than watching sheep, and, when visiting Grantham, reading books in the apothecary's house rather than trying to sell the farm's produce. Fortunately his maternal uncle, the rector of Burton Coggles about two miles from Woolsthorpe, and his discriminating schoolmaster, together persuaded his mother that farming was the wrong career for him, and he was allowed to return to school to prepare for university.

In June 1661 he was entered as a subsizar[†] at Trinity College, Cambridge, the college where his uncle had been educated. In his first two years he followed the traditional course dominated by Aristotelian philosophy, but in his third and fourth years he was branching out, reading Descartes' *Geometrie* and *Principia Philosophiae*, Kepler's *Dioptrice*, Galileo's *Dialogue Concerning the Two Chief World Systems* and *Siderius nuncius*, and Gassendi's account of Copernican astronomy. He also began working on mathematical problems. In the summer of 1665, the university was closed because of the plague, and Newton spent much of that and the following year in Lincolnshire, though often returning to Cambridge to use the library.[32] During the two plague years his intellectual activity was astonishing. He laid the foundations for his work on mathematics (including his method of 'fluxions', which was the basis of calculus), on optics (including the discovery that white light could be split into about seven colours), and on mechanics (including celestial mechanics). It is, of course, his work on mechanics that is relevant to this chapter and that is worth looking at in more detail.

[†] Sizars and subsizars were poor students who received a reduction in fees in return for menial services. It is not clear why Isaac's mother, who was relatively wealthy, should have agreed to this arrangement. She may have been irritated by his reluctance to farm; alternatively, the arrangement may have been initiated by Humphrey Babington, then a Fellow of the College, whose sister was the wife of the apothecary at Grantham, and who knew and liked Isaac.

By 1665 the battle between the Copernican system and the Ptolemaic system was long over. There was, though, still the problem that, though Kepler's work described how the planets move, it didn't explain why they move in that way. It was tempting to think of the Sun's rays, or of some other kind of 'spokes' from the Sun, acting as a sweeping broom but it didn't seem likely, and it wasn't easy to reconcile with Kepler's third law. And though Galileo's work gave a satisfactory account of terrestrial mechanics, the application of his work to celestial problems was not straightforward. Because inertia tends to maintain motion in a straight line, Descartes accounted for the motion of planets round their orbits by assuming that they were driven round by vortices in an all-pervading ether. Another confusing and unsatisfactory solution to the problem of planetary orbits was the misguided notion of 'circular inertia'. If in Galileo's experiments in which a ball continues to travel at constant speed along a flat frictionless surface, one thinks of the frictionless surface not as a perfect plane but as a smooth surface parallel to the circumference of the Earth, it seems intuitively right to suppose that the ball would carry on rolling right round the circumference, and if nothing altered it would continue to roll forever. The argument is sound but the explanation would not be 'circular inertia' but a combination of rectilinear inertia and the gravitational attraction of the Earth on the ball—precisely the combination that keeps satellites in their orbits.

Around 1665 Newton, seeking an explanation for planetary motion, became interested in the force necessary to constrain a body to move at a constant speed round a circular path. From his knowledge of inertia and of what would later become known as his own second law of motion, he was able to show that the force had to be directed towards the centre of the circle, and that its magnitude had to be proportional to the square of the speed, and inversely proportional to the radius. But Kepler's third law, obtained by comparing the orbital periods and distances from the Sun of the various planets, shows that for these planets the mean distances from the Sun and the orbital periods (and therefore the speeds) are not independent variables. The law is that the square of the orbital period is proportional to the cube of the planet's mean distance from the Sun. For an orbit that is close to circular, the planet's speed is nearly constant and the orbital period is inversely proportional to that speed. Bringing all these statements together and writing a few lines of algebra,[33] Newton came to a startling conclusion. If the gravitational force that the Sun exerts on a given planet is inversely proportional to the square of the distance between the centre of the Sun and the centre of the planet, then gravitational attraction can provide the forces necessary to account for the movements of the different planets round their orbits. This conclusion was reached assuming that the planets' orbits were close to circular, but about sixteen years later he proved that the same conclusion was true for more elliptical orbits.

Many years later the aged Newton, drinking tea under some apple trees in a garden in Kensington, told his old and close friend Dr Stukeley that:

> He was in just the same situation, as when formerly, the notion of gravitation came into his mind. It was occasion'd by the fall of an apple, as he sat in a contemplative mood.[34]

A fuller account of what is presumably the same event, is given by John Conduitt, the husband of Newton's niece, and his assistant when Newton was Master of the Mint:

> In the year 1666 he retired again from Cambridge...to his mother in Lincolnshire & while musing in a garden [probably at Woolsthorpe Manor] it came into his thought that the power of

> gravity (which brought an apple from a tree to the ground) was not limited to a certain distance from the Earth, but that this power must extend much further than was usually thought. Why not as high as the Moon thought he to himself & that if so, that must influence her motion and perhaps retain her in her orbit, whereupon he fell a-calculating what would be the effect of that supposition...[35]

Attractive though these stories are, it is difficult to believe that it was the fall of an apple that first suggested to Newton that the Earth's gravitational force might extend as far as the Moon. Kepler had long ago suggested a role for the Moon in causing tides, and if the Moon's gravitational pull extended to the Earth it would be odd to suppose that the Earth's didn't extend to the Moon. It is quite possible, though, that seeing the apple drop reminded Newton of the possible role of the Earth's gravity in maintaining the Moon's orbit, and that he then realized that his discovery of the inverse square law of gravitation made that hypothesis testable.

Since the distance of the Moon from the centre of the Earth was known to be about sixty times as great as the distance of the apple from the centre of the Earth,[36] he argued that the strength of the Earth's gravity on the Moon should be about 1/3600 times its strength on the Earth's surface. The crucial question then was: is this the strength that would keep the moon in its orbit round the Earth?[‡] The answer seemed to be that it was about 11% too low—probably because he used an

[‡] Newton did not know the mass of the Moon, but he did not need to. This is because, though a moon with, say, twice the mass would be pulled twice as hard towards the Earth, the acceleration would be the same because of Newton's second law, which states that the acceleration of an object produced by a net force is directly proportional to the strength of the force and inversely proportional to the mass of the object. It seems, then, that Newton must have been thinking about what was to become his second law as early as 1666.

inaccurate figure for the radius of the Earth. And though, some time later, when asked if the strength came out right he replied 'pretty nearly', he obviously was not happy with this result, since he dropped the subject for twenty years, and for a time was inclined to think that as well as responding to gravity the Moon might be carried along in an ethereal vortex, as Descartes had suggested.

Although Newton began thinking about what he called 'the lawes of Motion' in the 1660s, it was not until 1685, when he was in his early 40s and writing his *Principia*, that he worked out the final form of those laws. Having hesitated for two decades about the principle of inertia he finally adopted the notion of rectilinear inertia discussed both by Galileo and Descartes. His **first law** (often known as the law of inertia), is, in modern English wording:

> An object remains in a state of rest or of uniform motion in a straight line unless acted on by an unbalanced force.

His **second law** was also based on ideas that he had begun to explore twenty years earlier. It is:

> The acceleration of an object produced by a net force is directly proportional to the magnitude of the net force, and inversely proportional to the mass of the object.

Here acceleration means a change in velocity, and velocity means speed in a given direction. A change in velocity can therefore be a change in speed or a change in direction or both. The Moon going round its almost circular orbit is going at a steady speed, but it is continually changing its direction towards the Earth; it has a steady acceleration towards the Earth, corresponding to the steady gravitational force pulling it towards the Earth. Mars following a more elliptical orbit with the Sun at one focus of the ellipse, changes in both speed and direction of motion as the gravitational force acting on it varies.

Newton's second law accounts nicely for Galileo's law of free fall. Because the mass of the falling body doesn't change, and the force of gravity on it doesn't change significantly, the acceleration is constant.

Newton's **third law** also had its origin in work done in 1666, when Newton was interested in what happens when two objects collide. It is usually stated more succinctly than intelligibly:

> To every action there is always opposed an equal and opposite reaction.

More informatively:

> When two objects interact, the magnitude of the force acting on the first object equals the magnitude of the force acting on the second object, and the direction of the force acting on the first object is in the opposite direction to the force acting on the second object. The opposing forces act along the same line of action, and are of the same type—for example, contact, frictional, gravitational, or magnetic.

When a bat hits a ball, the force of the bat on the ball has the same magnitude as the force of the ball on the bat, but the two forces are in opposite directions. (Because the bat has a much greater mass than the ball, it follows from Newton's second law that the ball will accelerate much more than the bat.)

The third law does not only apply to collisions. Sitting in my chair, I exert a downward force on the chair equal to the upward force that the chair exerts on me. And it does not only apply to objects in contact. The attraction of the Earth for the apple is the same as the attraction of the apple for the Earth.

More generally: for any two bodies the force of gravitational attraction between them is directly proportional to the product

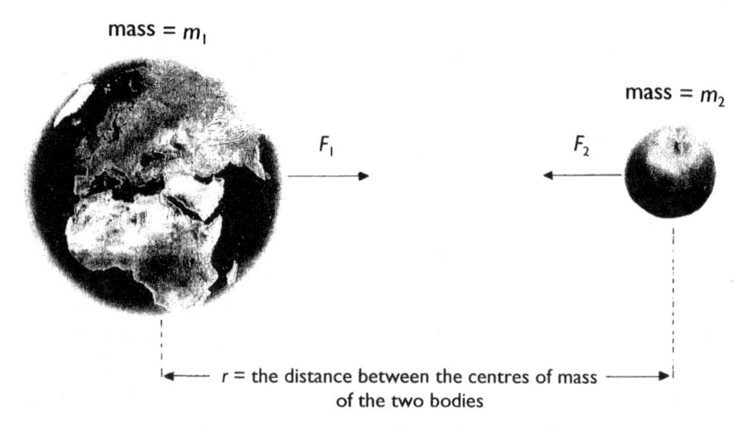

FIG 16 Diagram illustrating Newton's law of universal gravitation.

of their masses, and inversely proportional to the square of the distance between them—or, more strictly, between what we now call their centres of mass (see Figure 16). By the end of 1685, Newton had shown that, taken together, his three laws of motion and the inverse square law of gravitation could account for Galileo's observations of terrestrial motion and Kepler's observations of celestial motion. He had also returned to his 1666 attempt to correlate the Moon's orbit with the estimated strength of the Earth's gravity on the Moon. By this time, a more accurate figure for the Earth's radius was available and instead of an 11% discrepancy he found almost exact agreement. The Moon's inertia and the Earth's gravitational attraction together accounted for the Moon's movement. (Not only were Descartes' ethereal vortices unnecessary but, Newton showed, they were incompatible with Kepler's laws.)

The correlation of the Moon's orbit with the estimated strength of the Earth's gravity on the Moon did not itself prove that gravitation is universal—that is to say: any two bodies anywhere would attract one another with a force directly proportional to the product of their masses and inversely proportional to the square of the distance between them. But in Book III of his *Principia*, Newton showed that what he called 'the law of universal gravitation' could, along with his laws of motion, account for a great variety of observations: not just the straightforward motions of planets and their satellites, but also the interplanetary perturbation seen when the massive planets Jupiter and Saturn are in near conjunction; the perturbing action of the Sun on the motion of the Moon; the motions of comets; the phenomena of tides; the fact that the Earth is not quite spherical (bulging at the Equator and flattened at the Poles); and—a consequence of this bulging—the precession of the equinoxes.

This is the name given to the very small but progressive changes in the timing of the spring and autumn equinoxes, which we now know is caused by the gyration of the Earth's axis with an angular velocity that is so slow that it takes nearly 26,000 years to complete one revolution. It was the Greek mathematician Hipparchus, making a star catalogue in Rhodes in about 130 BC, who noticed that the position of a star called Spica at the time of the autumnal equinox was about 2° of celestial longitude different from the position of the same star observed by Alexandrian astronomers at the autumnal equinox about a century and a half earlier. Hipparchus attributed the slight shift to a very slow movement of the whole shell of fixed stars; sixteen centuries later Copernicus pointed out that if the heliocentric theory of planetary movement is right, the slight shift in the timing of the equinoxes could be explained by a very slow gyration of the Earth's axis. But why should the Earth's axis gyrate?

It was Newton who answered that question. Figure 17 shows the three kinds of movement of the Earth that need to be considered if the Earth's axis does gyrate. In the figure, the Earth is

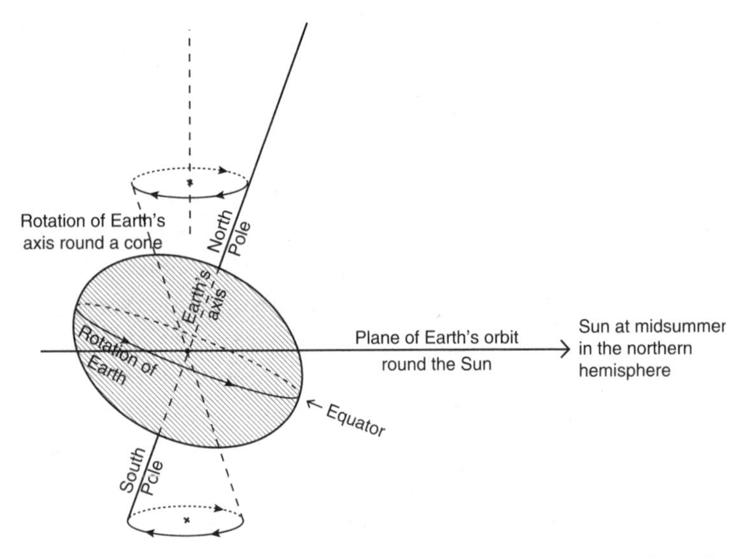

FIG 17 Diagram showing the three kinds of motion of the Earth: daily rotation, annual orbit, and gyration of axis.

shown hatched, and the bulging of the Earth at the Equator and flattening at the Poles is greatly exaggerated. Looked at from the North Pole, the Earth rotates anticlockwise daily. The axis of rotation is not perpendicular to the plane of the Earth's annual orbit round the Sun, but is tilted about 23° from the perpendicular. The direction of tilt remains virtually the same while the Earth goes round its orbit, and it is this that causes the seasons. But though the tilt remains the same, the change in the Earth's position relative to the Sun means that the tilt (of the northern half of the axis) is *towards* the Sun when it is summer in the northern hemisphere, *away* from the Sun when it is winter, and *neither* at the spring and autumn equinoxes. While the tilt of the axis is either towards or away from the Sun, the bulging of the Earth near the Equator causes small differences between the gravitational pull of the Sun on the halves of the Earth above and below the orbital plane, and these differences cause a slight torque—or turning force—tending to reduce the tilt of the axis. The gravitational pull of the Moon also leads to torques. If the Earth were not spinning, these torques would simply alter the tilt. But the Earth is spinning. The effect of the torques is, therefore, like the effect of attempting to alter the tilt of a spinning top or a gyroscope; it causes the spinning object to gyrate—its axis moving very slowly round, as if tracing out a cone (or two cones), in a direction opposite to that of the spin. It is this slow movement of the Earth's axis—taking nearly 26,000 years for a complete revolution—that causes the small but progressive and regular changes in the timing of the spring and autumn equinoxes.

The simple elegance and the staggeringly wide-ranging explanatory power of Newton's law of universal gravitation and his three laws of motion must have given him great satisfaction, but there was one aspect of the work that made him deeply unhappy. In a letter to Richard Bentley, the Master of Trinity College, written in February 1693 he explains why:

> Tis unconceivable that inanimate brute matter should (without ye mediation of something else wch is not material) operate upon & affect other matter wthout mutual contact... And this is one reason why I desired you would not ascribe innate gravity to me. That gravity should be innate inherent & essential to matter so y^{t} one body may act upon another at a distance through a vacuum wthout the mediation of anything else by & through wch their action or force may be conveyed from one to another is to me so great an absurdity that I believe no man who has in philosophical matters any competent faculty of thinking can ever fall into it. Gravity must be caused by an agent acting constantly according to certain laws, but whether this agent be material or immaterial is a question I have left to ye consideration of my readers.[37]

Fortunately, with the passage of time, physicists have become entirely happy with the concept of action at a distance, whether they are dealing with gravitational, electrical, or magnetic forces. As a character in Marmion's *The Antiquary* said, 'Familiarity begets boldness.'

4

SO WHAT IS HEAT?

Heat itself, its essence and quiddity, is Motion and nothing else.

Francis Bacon, 1561–1626[38]

Francis Bacon would eventually be proved right, but by the time steam engines had been invented, and engineers had become interested in the nature of heat, the prevailing view was that heat was a fluid. When Lavoisier was establishing the nature of combustion, in the 1770s, he referred to heat as *fluide igné* or *matière du feu* or, later, *caloric*. Whatever its name, he thought of it as a self-repellent, weightless fluid that raised the temperature of any body to which it was added, and lowered the temperature of any body from which it was removed. The self-repellent quality explained why increasing the amount of caloric in the interstices of a body caused it to expand; the weightlessness explained why heating a body did not generally change its weight. The theory that caloric is a substance would also explain why, when heat is exchanged between two or more bodies, the total amount lost is equal to the total amount gained. And finally, with some ingenuity, the theory provided Lavoisier with an explanation of the puzzling fact, noticed by Joseph Black in Glasgow, that when a gently stirred mixture of ice and ice-cold water is warmed, the ice gradually turns to water but, until all the ice has melted, both the ice and the water remain at the same temperature (0°C). Lavoisier's interpretation was that the added caloric and the ice combine to form a kind of chemical compound—liquid water—so that the properties of the caloric are suppressed and there is no change in temperature.*

This 'material' (and for a time highly successful) theory of heat was destroyed, over the next seventy years, largely by the efforts of three men, of whom two did elegant experiments and the third produced an elegant argument based on the results of experiments by others. Their work not only made the notion of caloric untenable but also provided the basis for the discovery of the law of conservation of energy, with heat as one form of energy.

The first of the two experimentalists, now generally known as Count Rumford, had one of the most extraordinary lives of any famous scientist, forever switching from one activity to another, and with a habit of solving practical problems in ways that shed light on the underlying theory.[39] He was born Benjamin Thompson, in 1753, to a farming family in Woburn, Massachusetts. At school he showed an interest in mathematics and mechanical devices, and though he left school at 13, at 16 he was fascinated by a long chapter on 'fire' in a book on chemistry

* The combination of caloric with other substances to form chemical compounds had already been invoked by Lavoisier to explain the generation of heat by combustion. What happens in combustion, he said, is that the flammable substance already contains *combined* caloric, but that oxygen—the gas then recently discovered by Scheele, in Sweden, and by Priestley, in England—has a higher affinity than caloric for the flammable substance and so displaces it; it was the freeing of previously bound caloric, he believed (of course wrongly), that accounted for the heat produced by combustion.

written by the Dutch physician and chemist Hermann Boerhaave. Boerhaave's notion of heat was a curious hybrid between the 'motion school' and the 'material school'; it was, he thought, *the vibration of particles of ordinary matter* that was responsible for the phenomena of heat, but what caused those vibrations was the motion of very small, solid, but weightless, particles of 'fire'. Thompson's fascination with heat was to resurface at intervals in the course of his eventful life.

Self-education dominated his early career. Apprenticeships in trading stores were soon seen to be unsuitable, and he broadened an abortive experience as a medical apprentice by going to various scientific lectures. His obvious abilities led first to tutoring, and then to an invitation to teach in a school in Rumford (later Concord), New Hampshire. The invitation came from a clergyman whose 33-year-old daughter, married to a wealthy landowner, had recently been widowed; within four months of his arrival in Concord, Thompson, still only 19, married her—or as he later put it was married to her. Active or passive, the marriage gave him both wealth and access to the political and social circle surrounding the royalist Governor, the first of a series of contacts with influential people that would continue throughout his life. The following year, the Governor made him a major in the second provincial regiment of New Hampshire—to the disgust of more experienced officers—and he started a long-term technical interest in gunpowder.

In 1774, a year before the start of the American War of Independence, Thompson's strong support for the established order put him in a difficult position, and in December of that year he was charged by a citizens' vigilante committee of Concord and nearby towns with being hostile to American freedom. Although the evidence was judged inadequate to secure a conviction, he fled to Boston, leaving his wife and year-old daughter in New

Hampshire. He never saw his wife again, and he didn't see his daughter for more than twenty years.

In Boston, he acted as a spy for the Governor, using a sophisticated invisible ink made from oak galls, having an affair with the wife of the printer of a rabidly revolutionary newspaper, and liaising with a trusted member of the (revolutionary) Provincial Congress, who was at the same time a paid informer for the British. When, in 1776, the British army withdrew from Boston, Thompson sailed for England in the frigate carrying despatches to the Secretary of State for the Colonies, Lord George Germain, in London. Before leaving America, Thomson had armed himself with letters of recommendation from the royalist Governor of New Hampshire and from the general in charge of British forces, and in London he presented himself and his letters to Germain.

For Germain, who had been appointed to the post only the previous year and who was woefully ignorant of the American scene, the arrival of a knowledgeable American was a godsend. Thompson was immediately employed as a Private Secretary in the Colonial Office, though he also found time to pursue his researches on gunpowder, sending a paper to the Royal Society, getting to know Sir Joseph Banks (the President) and in 1779 being elected a Fellow. A year later, he was made Undersecretary of State for the Colonies, but during the following months he felt he was being blamed for Germain's incompetence, and he was himself suspected of selling British naval secrets to a French spy on trial in London. Germain's tenure was insecure and there was no prospect that Thompson's job would outlast it. A military career seemed a possible solution to his problems, and in the summer of 1781, having been appointed a Lieutenant Colonel in charge of the King's American Dragoons (at a cost to himself of £4,500), he set sail for America. Here he was involved in skirmishes at Charleston, but Cornwallis' surrender to Washington at Yorktown effectively ended active campaigning. After Charleston was evacuated, Thompson camped with his regiment at Huntingdon, Long Island, and ensured his own infamy in America by pulling down a Presbyterian church to build fortifications, burning apple trees for fuel, and using gravestones to make baking ovens. When the War of Independence ended he returned to England, where grossly exaggerated reports of his achievements enabled him to retire as a full colonel on half pay. There was, though, no prospect of resuming his diplomatic career, so after having his portrait painted by Gainsborough he moved to continental Europe.

Thompson had a gift for meeting people. In the cross-channel boat he met the historian Gibbon, who later referred to him as 'Mr Secretary-Colonel-Admiral-Philosopher Thompson.' In Strasburg, wearing his colonel's uniform and sitting on a fine horse, he attended a French military parade commanded by Prince Maximilian de Deux Ponts, who was both a French Field Marshal and a nephew of Karl Theodor, the Elector of Bavaria. The Prince was intrigued by the handsome 30-year-old English colonel and not only introduced him to members of the parading regiment who had recently served in the American War (on the opposite side), but also encouraged him to visit Karl Theodor in Munich and gave him a warm letter of introduction. Less than six months after leaving England, Thompson had been offered a job at the Elector's court—to advise on ways of reorganizing and modernizing the Bavarian Army.

As a British colonel, Thompson needed permission before he could serve in a foreign court, but before returning to Britain to seek it he spent some time in Vienna, having promised to report any matters of scientific interest to his friends at the Royal Society. Back in London, he not only got permission to accept the job in Munich, but was also knighted. How much this was a reward for past services, how much an attempt to improve relations with

Bavaria in order to lessen the influence of the French, and how much a payment for expected future spying (which he had discussed with the British ambassador in Vienna) is not clear.

By 1784, Sir Benjamin Thompson was established in Munich where for the next four years he studied the problems of the Bavarian army, improved his German and French, and had affairs with two sisters, both countesses, one (with whom he later had a daughter) the former mistress of the Elector, but now fat; the other, slim, intelligent, and happy to help him with his writing. In 1788 he produced a plan, acceptable to the Elector. His general aim was 'to unite the interest of the soldier with the interest of civil society, and to render the military force, even in time of peace, subservient to the *public good*';[40] and he did in fact make the army both much more efficient and much more humane. He established a military academy and a veterinary school, and—significantly for his later work—he reorganized the manufacture of cannon.

Particularly remarkable was his success in reclothing the army, where his interest in heat surfaced in his study of the thermal conductivity of substances of different kinds. He compared the ability to conduct heat of various materials used in clothing, by embedding the bulb of a mercury thermometer in a globe filled in turn with silk, wool, cotton, linen, fine fur, and eider down, and seeing how fast the thermometer could respond to changes in the outside temperature.[41] The results suggested that it was the amount of air trapped within a fabric that most affected its conductivity; and he realized that it was the trapping of air in the interstices of the fabric so that it could not form what would later be called convection currents that accounted for the insulating quality of the fabric.

Direct evidence for convection currents came some years later from an ingenious and elegant interpretation of a chance observation that he made while doing an experiment.[42] He was using an unusual thermometer, which had a large copper bulb, more than four inches in diameter, attached to a glass tube, and filled with spirits of wine (ethyl alcohol). Having heated the thermometer 'in as great a heat as it was capable of supporting' he 'placed it in a window, where the sun happened to be shining, to cool.' Glancing at the tube, which was vertical and 'which was quite naked (the divisions of its scale being marked in the glass with a diamond)' he 'saw the whole mass of the liquid in the tube in a most rapid motion, running swiftly in two opposite directions, *up* and *down* at the same time.' The movement, he explains, was visible because the alcohol contained fine particles of dust that had accumulated in the copper bulb, and which 'being illuminated by the sun's beams, became perfectly visible (as the dust in the air of a darkened room is illuminated and rendered visible by the sunbeams which come in through a hole in the window shutter).' The internal diameter of the tube was just over a tenth of an inch, but on examining the tube with a lens he saw 'that the ascending current occupied the *axis of the tube*, and that it descended by *the sides of the tube*.' However, 'on inclining the tube a little, the *rising* current moved out of the axis and occupied that side of the tube which was uppermost, while the *descending* current occupied the whole of the lower side of it.' This is just what would be expected if the flow were the result of differences in density of the fluid caused by differences in temperature—in other words, if the currents were convection currents.

His discovery of convection currents would later be rewarded by the Copley medal of the Royal Society, but his immediate aim was to make use of his findings to improve the clothing of the Bavarian army. Finding that existing clothing manufacturers were reluctant to cooperate, he conscripted beggars—about 4% of the population of Munich at that time[43]—and put them, and their women and children over five years old, in military workhouses, promising them training, wages, warm dinners, and medical care. The working day was twelve to fourteen hours, and there were two hours of schooling for children. He explained the principle on which the military workhouse was based:

> To make vicious and abandoned people happy, it has generally been supposed necessary, *first*, to make them virtuous. But why not reverse this order? Why not make them first *happy*, and then virtuous?[44]

Having so many people to look after, he put great effort into improving artificial lighting, fireplaces, and stoves; he encouraged soldiers to grow their own vegetables, and he made the potato popular in Bavaria. He also persuaded the Elector to convert a neglected marshy deer park into a garden, based on London's Kew Gardens including the Chinese pagoda; known as 'The English Garden' it still exists and is the largest park in Munich.

Thompson's various successes led to his appointment as Minister of War, Minister of Police, and Court Chamberlain; and in 1792 he became a Count of the Holy Roman Empire, adopting the name Rumford after the town where his career had started. But 1792 was not a happy time for the new Count. The outbreak of war between France and Austria, and later Prussia, made the aging Elector's policy of appeasement unpopular, and Rumford himself had bouts of sickness. In March of the following year, he took sick leave and spent nearly sixteen months in Italy, meeting Volta (who had not yet invented the first electric battery), experimenting on complementary colours and the thermal conductivity of steam, advising hospitals about the rebuilding of kitchens, spending time with his mistresses, and meeting and becoming an intimate friend of Lady Palmerston, the mother of the future British Prime Minister. Returning to Munich in

July 1794, he received a warm welcome, but the situation had not improved; fifteen months later, anxious to publish his scientific, technical, and social work, he obtained leave to spend time in Britain.

His arrival began disastrously, with the theft of a large trunk full of his papers, but the nine months he spent in Britain were both busy and happy. Renewing his acquaintance with Lady Palmerston, he was asked to solve the problems associated with her fireplace in Hanover Square, and in this he was so successful that during his stay in England he is said to have modified 250 fireplaces in London alone; cartoons of Rumford fireplaces were drawn by both Gillray and Cruickshank, and in Jane Austen's *Northanger Abbey* the heroine finds that the 'fireplace where she had expected the ample width and ponderous carving of former times, was contracted to a Rumford.' All that was more than two centuries ago, but Rumford fireplaces are advertised on *Google* today.

The trouble with eighteenth century fireplaces was that too much of the heat went up the chimney and too much of the smoke came into the room. Rumford had elegant solutions to both these problems.[45] He realized that, with an open fireplace, most of the heat transferred to air by convection will be lost up the chimney. In contrast, a large fraction of the radiant heat produced by the fire will enter the room. By having smaller but more intense fires in much less deep fireplaces he could increase the proportion of heat liberated as radiant heat, and by making the sides of the fireplace slope (see Figure 18) he could ensure that a greater fraction of that radiant heat would get into the room rather than simply warming the side walls of the fireplace.

The problem of smoke was, he pointed out, largely a problem of the chimney and its 'throat' or junction with the fireplace. Traditional chimneys were very broad and with wide and deep

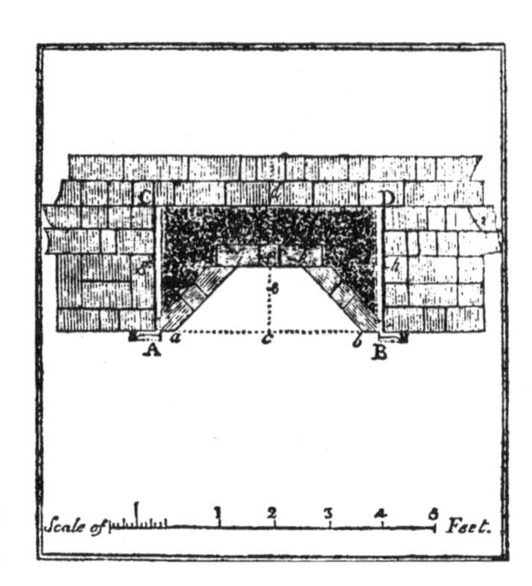

FIG 18 Plan of fireplace modified by Rumford.

throats, and the result of this was that the column of less dense, hot, and smoke-laden air passing upwards in the centre of the chimney was cooled where it was in contact with the colder denser air around it, leading to turbulence and the movement of some cool smoky air down the chimney into the room. To prevent this, he suggested three things. Firstly, the throat should be no wider (side to side) than the width of the back of the fireplace and no deeper (from front to back) than about four inches—see Figure 19. Secondly, the bottom of the throat should not be too high above the fire, so that the column of rising air will still be very hot as it enters the throat. And thirdly, what he calls the breast of the chimney—the back part of the mantle—should be rounded so that air may flow smoothly under the mantle to 'unite quietly with the ascending column of smoke.' Together, these changes should ensure that there is a continuous vigorous upward flow of air through the throat, so that any turbulence higher in the chimney will have no effect on the room containing the fireplace. He pointed out that in converting existing fireplaces, if the

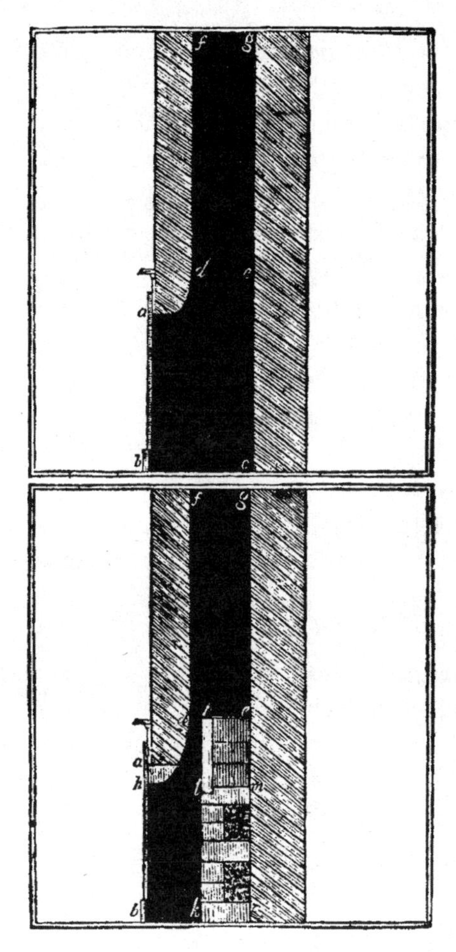

FIG 19 Rumford's modification of a chimney. The lower picture shows the decreased depth of the fireplace, the narrowing and lowering of the throat, and the rounding of the back part of the mantel.

wall of bricks or stone built at the back of the fireplace to make it more shallow is continued upwards into the throat of the chimney, it can serve also to reduce the depth of the throat, though the top part of the wall will need to be removable to allow access to chimneysweeps.

As well as being so successful in London, Rumford was welcomed in Dublin, Edinburgh, Bath, and Harrogate, and his 22-year-old daughter, whom he had not seen since she was one, came to visit him. Feeling that his career in Bavaria was not too secure, he began to consider possible futures in England or America, and in 1796 he offered substantial sums to the Royal Society, in London, and the American Academy of Arts and Sciences, in Boston. The money was to be used to create Rumford medals to be awarded each year to the authors of 'the most Important Discovery or useful Improvement...on *Heat* or on *Light*,' which should 'tend most to promote the good of mankind.' With characteristic modesty he expected himself to be the first recipient, and in London he was not disappointed.

In late July 1796, he received an urgent summons to return to Munich to help defend the city from the Austrians and the French. He returned, taking his daughter with him, and using subtle diplomacy he was so successful that the Bavarians regarded him as a hero and showed their gratitude by making his daughter a countess. In this situation Rumford might well have rested on his laurels, but over the next year he did the work on heat that began to topple Lavoisier's theory of caloric, that would eventually lead to the discovery of the law of conservation of energy, and that, incidentally would, I hope, help to justify the space devoted to him in this book.

> Being engaged, lately, in superintending the boring of cannon, in the workshops of the military arsenal at Munich, I was struck with the very considerable degree of heat which a brass gun acquires, in a short time, in being bored; and with the still more intense heat (much greater than that of boiling water, as I found by experiment) of the metallic chips separated from it by the borer...
>
> From *whence comes* the heat actually produced in the mechanical operation above mentioned?

> Is it furnished by the metallic chips which are separated by the borer from the solid mass of metal?

The quotation is from the introduction to Rumford's paper to the Royal Society, read in January 1798.[46] The answer to the two questions depends on the nature of heat. If heat is 'an igneous fluid'—the caloric theory—then it must leak out or be squeezed out of the chips of metal. If heat is essentially motion, then it must be generated by friction. Rumford's way of answering the questions was, in outline, breathtakingly simple. By using a blunt borer you can arrange things so that you have as much friction but a much smaller quantity of metal chips. Will you get as much heat?

Figure 20 is taken from Rumford's paper. The top drawing shows a cannon as it comes from the foundry. In a footnote, Rumford explains that brass cannon are cast solid in the vertical position with the muzzle upwards, and about 60 cm longer than the desired final length.[47] The role of the extra length is to exert pressure on the metal in the lower part of the mould, during the time it is cooling, so that 'the gun may be the more compact in the neighbourhood of the muzzle; where, without this precaution, the metal would be apt to be porous, or full of honeycombs.' Normally, this extra length—the *verlorner Kopf* ('lost head' in English)—is cut off before boring starts, but Rumford realized that by leaving it attached to the cannon he could carry out his experiment with the blunt borer on the extra length, and would not have to sacrifice a good cannon. He was, after all, now a general in the Bavarian army as well as a scientist.

Because the newly cast cannon has to be shaped on the outside as well as having a hole bored down its middle, it is held horizontally in what is in effect a horse-driven lathe that rotates it at about one turn every two seconds. The lower drawing in Figure 20 shows

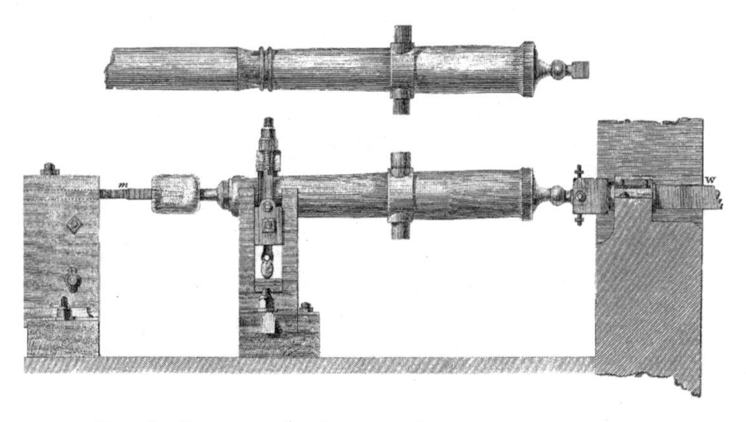

FIG 20 Rumford's cannon-boring experiments.

how Rumford shaped the *verlorner Kopf*, reducing it to a cylinder nearly 25 cm long and rather less in diameter, attached by a narrow neck to the rest of the cannon. Using a sharp steel borer, he then bored a hole a little over 9 cm wide along most of the length of the cylinder but stopping 7 cm from the end attached to the rest of the cannon. A narrow transverse hole was drilled into this end of the cylinder so that, when the cannon was not being rotated, a mercury thermometer could be inserted to measure the temperature. For the crucial experiment the cylinder was carefully wrapped in a coating of thick flannel (to minimize heat loss), and the sharp steel borer was replaced by what is described as a blunt steel borer, though 'blunt' is rather an understatement. It was a piece of hardened steel about 1.6 cm thick and nearly 9 cm wide, so the area of the grinding surface was nearly 15 sq. cm. Attached to an iron bar, this piece of steel was 'forcibly shoved (by means of a strong screw) against the [rotating] bottom of the bore of the cylinder' with a force equivalent to a weight of about 4500 kilograms.

At the end of thirty minutes (and nearly a thousand revolutions), the temperature of the cylinder had risen by nearly 39°C; and the flannel jacket must have been fairly effective since, with boring stopped, it took 34 minutes for the temperature of the cylinder to fall by 10 degrees. When Rumford removed the borer he found there was only about 54 grams of 'metallic dust, or rather scaly matter, which had been detached from the bottom of the cylinder.' Clearly, using a blunt borer greatly reduced the quantity of metal dust without having a similar effect on the amount of heat produced**—just the result predicted by the frictional theory and difficult to explain if heat was a fluid emerging from the metal.

Since the weight of the cylinder was nearly a thousand times the weight of the metallic dust, for the caloric explanation to be acceptable you would have to assume that the conversion of each *gram* of metal into dust liberated enough heat to raise the temperature of nearly a *kilogram* of metal by nearly 39°C. Another (and later) way of looking at the problem is this: the total amount of heat produced in this experiment was about 236,000 calories which would be enough to melt something like 1,700 grams of brass.[48] For the caloric theory to be tenable, then, the conversion of the 54 grams of brass into metallic dust must have liberated enough heat to melt about 1,700 grams of brass—that is, roughly thirty times its own weight. Whichever way you look at it, it seems that the caloric theory is tenable only if brass normally contains an improbably large amount of 'combined caloric', which is released when the metal is ground into dust.

When Rumford read his paper to the Royal Society in January 1798 he was 44 and at the peak of his scientific career. The rest

** The amount of heat produced could be calculated from the rise in temperature, knowing the mass of the cylinder, and the specific heat capacity of gunmetal.

of his life was filled with his characteristic mixture of scientific experiments, politics, and enterprise. Failing in his hopes of becoming Bavarian ambassador in London, he turned back to his enthusiasm for applied science. Appropriately, the man who had worked on warm clothing and efficient fireplaces became the founder of the Royal Institution, with its interest in 'the applications of science to the useful purposes of life.'[49] Rather less appropriately—and less successfully—he married the wealthy widow of Lavoisier, whose caloric theory he had demolished.

The second man whose experiments helped to establish the principle of the conservation of energy was James Prescott Joule, born in 1818.[50] Like Rumford, he never went to university and was interested in science, but otherwise their lives could scarcely have been more different. He was a delicate child, and his father, a wealthy brewer in Salford, near Manchester, was anxious about his education. He and an elder brother were therefore taught at home until he was 16, when they were sent for tutorials to the ageing John Dalton, at that time President of the Manchester Literary and Philosophical Society, and already famous for his advocacy of the atomic theory and for his study of colour blindness. Dalton taught them arithmetic and geometry, and had introduced them to chemistry, when he had a stroke—but not, apparently, before he had inspired the younger brother to a life in scientific research. By this time, the family home had been moved to the more congenial Swinton, near Manchester, and a room was set aside for Joule to use as a laboratory. It was in his 20s, and spending much of the day working in the brewery, that he did his most interesting scientific work.

More than forty years had passed since Rumford's cannon-boring experiment, and during that time Volta had invented his 'pile'[†] and discovered current electricity; the generation of electricity in that 'pile' and in other 'voltaic cells' had been shown to be the result of chemical changes; conversely, Faraday had shown that electric currents passing through aqueous solutions could cause chemical changes—a process he called *electrolysis.* Oersted had noticed that a wire carrying an electric current could deflect a compass needle, and Faraday had shown both that a current-carrying wire moved spontaneously in a magnetic field, and that moving a wire in a magnetic field generated an electric current in the wire. Finally, Ohm had shown that (at a constant temperature) the current passing through a conductor in an electrical circuit was directly proportional to the potential difference (or drop in voltage) between the ends of the conductor. It was against this broader background that Joule started his investigation of the nature of heat.

The elegance of Joule's work lay partly in his individual experiments and partly in the way he used a variety of those experiments to explore the quantitative connections between electrical, chemical, mechanical, and thermal effects. The results would show not only the interconvertibility of what we now call electrical energy, chemical energy, mechanical energy, and heat (thermal energy),

[†] In 1799 Volta found that when a disc of copper and a disc of zinc were separated by a cloth dampened with salt water, a small voltage could be detected between the two discs; and if they were connected by a wire a very small current was maintained. (Arrangements of this kind later became known as *voltaic cells.*) By making a 'pile' of pairs of discs arranged in this way (with the metals always in the same order) and connecting wires to the top copper plate and the bottom zinc plate he showed that there was a substantial voltage between the wires and that this could be used to drive a substantial current. (It is necessary to use salt water to dampen the cloth between the discs because pure water is only very slightly ionised into H^+ and OH^- ions, and is therefore a very poor conductor of electricity.)

but also that in any conversion of one form of energy into another the total energy remained the same. Energy was conserved.

He began by showing that the rate at which heat was produced by a steady electric current passing through a coil was proportional to the product of the resistance of the coil and the square of the current passing through it.[‡] In an experiment, reported in December 1840[51] when he was 22, he used a battery of voltaic cells to generate the current, and he measured the rate of heat production by keeping the coil in an insulated vessel containing gently stirred water, and measuring the change in temperature of the water with an extremely sensitive thermometer of his own design. The fact that with a known resistance and a known current it was possible to calculate the rate of heat production, suggested that there was some sort of equivalence between the amount of chemical change going on in the voltaic cells and the amount of heat produced in the coil. In modern terms we would say that the release of energy by the chemical changes going on in the voltaic cell was equivalent to the amount of heat produced in the coil.

Following up this idea, Joule did what with hindsight we can see was an elegant experiment, but one which was so novel and so unexpected that for some years it was almost totally ignored. Instead of passing current from the battery of voltaic cells through a coil he passed it through water,[52] some of which was thus decomposed (we should now say 'electrolysed') into hydrogen and oxygen—the amount of each being measured. The water was

‡ This makes sense because the current through the coil, the resistance of the coil, and the voltage drop across the coil, are related by Ohm's law (voltage drop = current × resistance). It follows that the product of the resistance and the square of the current is equal to the product of the voltage drop and the current—which product, we now know, is equal to the power of the electric current, i.e., the rate at which it delivers energy.

contained in an insulated container to prevent heat loss, so that by measuring the rise in the water temperature, the amount of heat gained by the water could be estimated. This amount was compared with the amount of heat gained during an *identical* period of time by an *identical* quantity of water containing a coil with a resistance chosen so that (with the same battery) *the flow of current through the coil was the same as the flow of current that had passed through the water.* It turned out that the amount of heat gained by the water, when it itself had carried (and to a small extent been decomposed by) the current, was substantially less than the amount of heat gained by the water when the same current had been carried by the immersed coil.[53] How was this difference to be explained? Joule's hypothesis was that what he called 'the lost heat' had been used up by the decomposition of some of the water to form hydrogen and oxygen:

$$\text{'lost heat'} + 2H_2O \rightarrow 2H_2 + O_2.$$

If that was right, reversing the reaction should lead to the production of a similar amount of heat, and in fact the amount of the 'lost heat' *was*, he calculated, similar to the amount of heat that would be formed by the combustion of a quantity of hydrogen equal to that which had been formed by the passage of the current. In modern language, we would say that the energy that didn't appear as heat, because it had been used to split water into hydrogen and oxygen, could be regained as heat by burning the hydrogen and so re-forming the water:

$$2H_2 + O_2 \rightarrow 2H_2O + \text{'heat of combustion'}.$$

It seemed, then, that where in the first experiment there had been an equivalence between the amount of chemical change in the voltaic cell and the amount of heat produced by the coil, in the second experiment there was an equivalence between the amount of chemical change in the voltaic cell, and the *sum* of the amount of heat produced and the amount of chemical change produced in the water. This kind of bookkeeping suggested that something was being conserved, but it didn't yet have a satisfactory name. In discussing his experiments Joule used terms such as 'calorific effect', 'chemical heat', 'free heat', or 'latent heat'.

In 1842, at about the same time as Joule was experimenting with voltaic cells, Julius Mayer, a young medical doctor in Heilbronn, then a quiet country town in Würtemberg, Germany, published a fairly short paper in *Liebig's Annalen der Chemie und Pharmacie*,[54] which was remarkable in several ways. Firstly, it was all about physics—which was odd coming from a country doctor and appearing in a journal of chemistry and pharmacy. Secondly, most of it was devoted to a discussion of the two meanings of the noun *force* (*Kraft* in German)—a word which at that time was sometimes used in the Newtonian sense of *a pull* or *a push*, and sometimes in the sense that is now conveyed by the word *energy*.[§] He suggested that different forces in the second sense (i.e., what we should now call different kinds of energy) are equivalent to one another; and he thought that what he called '*Fall Kraft*' (potential energy), 'motion' (kinetic energy), and heat could all be interconverted—and interconverted without any change in the total amount of energy. This was the earliest statement of the

[§] It was not until the early 1850s that William Thomson used the English word *energy* to mean 'the capacity to do mechanical work'.

principle of the conservation of energy. Thirdly, and no less remarkably, the final brief paragraph contained the earliest (though not very accurate) statement of the mechanical equivalent of heat:—'the fall of a weight from a height of about 365 metres corresponds to the heating of an equal mass of water from 0°C to 1°C.' (In modern units this is saying that 3.59 joules# of work are equivalent to one calorie.) Astonishingly, Mayer did not explain the experimental or theoretical basis of this statement.

It was not until 1845, three years later, that Mayer at last explained how he had arrived at his estimate.[55] It was the result of a simple, elegant, and convincing argument based on the observations of three different sets of French investigators, all published many years earlier. The crux of the argument is this. It is well known that *more* heat is needed to cause a specified rise in the temperature of a given mass of gas, if the gas is held at a constant pressure rather than at a constant volume. If one could assume that the extra heat is needed solely because the gas does mechanical work as it expands against the pressure, and the extra heat is equivalent to that work, then knowing how much extra heat is needed, and knowing the amount of expansion and the pressure, one could calculate the mechanical equivalent of heat. But is that assumption true? Might it be that, as the gas expands, its structure changes, and that the process of changing that structure itself requires work or does work? Fortunately, Mayer remembered that many years earlier Gay-Lussac had shown that when a gas expands into a vacuum there is no overall gain or loss of heat. That observation excludes the possibility that permanent

The 'joule'—named after James Prescott Joule—is the modern unit of work, one joule being the work done by a force of one newton moving one metre in the direction of the force. One newton is defined as the force required to give a mass of one kilogram an acceleration of one metre per second per second. (Happily, it is also roughly the gravitational force acting on a smallish apple.)

structural changes requiring or doing work are associated with the expansion.

Totally unaware of the important work of Mayer, Joule—having established equivalences between chemical, electrical, and thermal effects—decided to bring *mechanical* effects into the picture. Instead of using voltaic cells to generate current to pass through a coil, he rotated the coil itself in a strong magnetic field, so inducing a current and heating the coil.[56] By including a galvanometer—a device for measuring electric current in arbitrary units—in the circuit, he could also measure the strength of the induced current. He found that the rule governing the rate of heat production was just the same as in the earlier experiments using voltaic cells. He then arranged that the vessel containing the coil was rotated by a system of falling weights so that he could calculate just how much mechanical work was being done in a given time, and compare that figure with the amount of heat produced in that time. (By using a strong magnetic field and a coil with many turns, he could ensure that the work done against friction in the apparatus was negligible compared with the work done in generating the electric current. And by using heavy weights that could provide sufficient power while falling relatively slowly, he could ensure that there was no significant loss of kinetic energy when the weights hit the ground.) From a series of thirteen experiments of this kind he obtained answers which (in modern units) ranged from 3.15 to 5.59 joules per calorie, with an average value of 4.50 ± 1.21 joules per calorie. These figures were reported at the British Association Meeting in Cork in 1843. The wide scatter was the result of the many technical difficulties in the experiments but, given the scatter, the results were in reasonably good agreement with the accepted modern value of 4.18 joules per calorie. (They were also in reasonably good agreement with Mayer's estimate of 3.59 joules per calorie but Joule was still quite unaware of Mayer's work. It had been published the previous year but had caused very little stir.[##])

If a given amount of heat is equivalent to a given amount of mechanical work, estimates of the mechanical equivalent of heat should give the same answer irrespective of the kind of mechanical work done. To test this, Joule went on to do experiments that involved different kinds of work.

He first looked at the thermal effects of changing the volume of a given mass of gas.[57] When air was compressed, work was done on it and it got hotter. Calculation of the mechanical equivalent of heat gave a figure of 4.35 joules per calorie. When compressed air was allowed to expand against atmospheric pressure, work was done by the air and it got colder. Calculation of the mechanical equivalent of heat from experiments of this kind gave figures ranging from 4.08–4.42 joules per calorie. There was, though, a snag—the same snag that Mayer had had to cope with. The calculations are valid only if one can assume that the changes in volume of the gas had *not* been associated with changes in structure that either required work to be done or that did work. Mayer had recognized this problem and solved it by referring to the experiment by Gay-Lussac. Joule, too, recognized the problem—indeed, he had shown that when work is done in winding up a watch spring, no heat is produced: all the work is used to change the internal structure of the spring. But he didn't know about Gay-Lussac's experiment. He therefore did what was essentially a

[##] Even five years later, in 1847, Helmholtz's famous essay *Über die Erhaltung der Kraft* (*On the conservation of force*), which distinguished between *forces* in the Newtonian sense, which are not conserved, and *forces* that are conserved, made no reference to Mayer's work.

more accurate version of that experiment. He had two identical fixed-volume copper vessels connected by a stopcock. With the stopcock closed, one was filled with compressed air, the other was evacuated. The pair of vessels were placed in a calorimeter—that is, a well insulated vessel containing water and fitted with a sensitive thermometer—and left until the reading on the thermometer was steady. The stopcock was then opened, allowing air to escape from the vessel containing compressed air to the vessel that contained a vacuum. The water was gently stirred and its temperature taken. As Joule expected, no temperature change could be detected. If the doubling in volume of the gas that occurred when the stopcock was opened involved a change in the structure of the gas that either required or did work, there would necessarily have been a change in temperature.

Further experiments were designed to measure the heat produced by friction in liquids. Falling weights were used to drive paddle wheels immersed in water, sperm oil, or mercury, the liquid being contained in a well-insulated vessel fitted with an accurate and sensitive thermometer (see Figure 21).[58] In three experiments of this kind, he obtained estimates of the mechanical equivalent of heat, in the range 4.15–4.17 joules per calorie. Yet another set of experiments, in which a pair of cast-iron rings in a calorimeter filled with mercury were made to rub against each other by falling weights, gave estimates in the same range.

Curiously, Joule's early experiments with voltaic cells were almost completely ignored. The Royal Society had not been interested even in the highly original paper 'On the heat evolved during the electrolysis of water'. The thirteen scattered estimates of the mechanical equivalent of heat, reported at the British Association Meeting in Cork in 1843, had more publicity but were criticized because of their scatter. In 1847, though, Joule described his paddle-wheel experiments at the Oxford meeting of the British

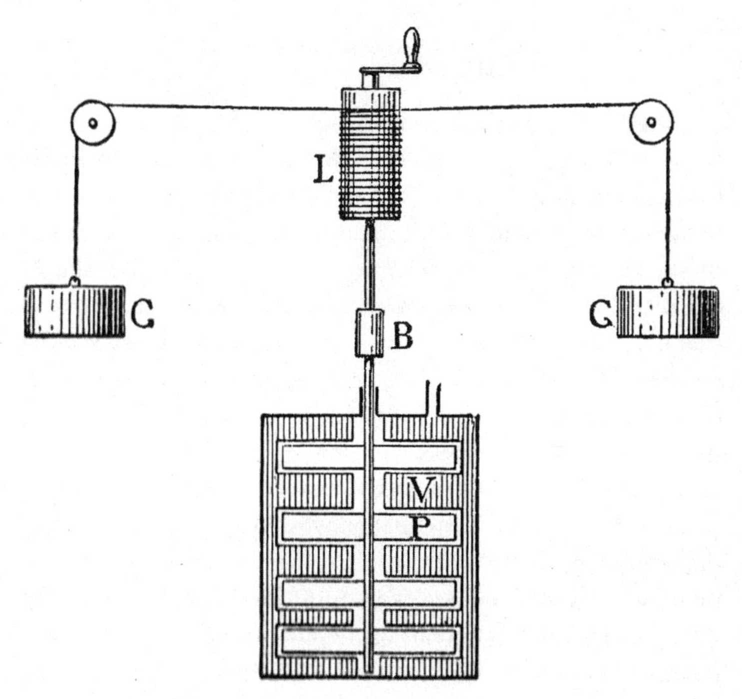

FIG 21 Joule's determination of the mechanical equivalent of heat. Falling weights rotate a paddle wheel whose paddles (P) turn in the narrow space between vanes (V) fixed to the wall of the vessel.

Association, and despite being given only enough time to make a brief presentation, he succeeded in arousing the immediate interest of both William Thomson (later Lord Kelvin) and Michael Faraday. The following year he was elected a corresponding member of the Royal Academy of Sciences at Turin; in 1850, in his early 30s, he was elected a Fellow of the Royal Society; and in 1870, seventy-eight years after Rumford, he was awarded the Society's Copley medal. But the most lasting sign of the recognition of his work was the adoption of the 'joule' as the international unit of both work and energy.

A TAILPIECE TO CHAPTER 4

The work of Rumford, Mayer, and Joule firmly established the *law of conservation of energy*, now also known as the *first law of thermodynamics*. That law places no limits on the transformation of energy in any one form into energy in any other form. Yet experience shows that there is an asymmetry. The *complete* (or almost complete) transformation of mechanical energy or electrical energy or chemical energy into heat is easily achieved. The *complete* transformation of heat into mechanical energy or electrical energy or chemical energy is not possible. Understanding this asymmetry led to the second law of thermodynamics, with very far-reaching consequences. To explain these consequences satisfactorily requires more mathematics than is appropriate for this book, but the first (and crucial) step in understanding the asymmetry came from one of the most elegant and far-reaching thought experiments in the history of science; and that step requires only the simplest mathematics. It is, though, not the easiest argument to follow, so I have put it in an appendix—see page 235.

5

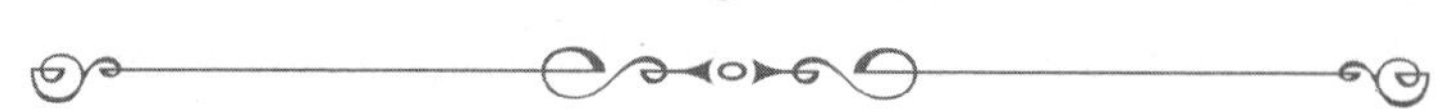

ELEGANCE AND ELECTRICITY

For a succession of simple elegant experiments leading to conclusions of great theoretical and practical importance, there is no better example than Michael Faraday's studies on electricity. But Faraday's studies are concerned only with current electricity, and in 1747, just over fifty years before current electricity was discovered, Benjamin Franklin—still a professional printer, though owning his own newspaper—did some experiments on static electricity, which are equally remarkable for their simplicity and their importance.

Five years before his famous and dangerous kite-flying experiment, Franklin was puzzled by some startling findings of Monsieur C.F. Du Fay, the Director of the Jardin de Roi in Paris, which seemed to show that electricity could be of two kinds.[59] A piece of gold leaf that had been charged with electricity by being touched by a rubbed glass tube was subsequently repelled by the tube. Similarly, a piece of gold leaf that had been touched by a rubbed amber rod was repelled by the rod. But what Du Fay found, to his great surprise, was that a piece of gold leaf that had been touched by a rubbed glass tube was subsequently *attracted* by a rubbed amber rod; and a piece of gold leaf that had been touched by a rubbed amber rod was subsequently *attracted* by a rubbed glass tube. He concluded that the electricity found on rubbed glass, which he called *vitreous*, was different from the electricity found on rubbed amber, which he called *resinous*. It seemed that objects bearing the same kind of electricity repelled each other, whereas objects bearing opposite kinds attracted each other.

Franklin was aware that if a man stood on an insulating sheet of wax, and his knuckle was repeatedly touched with a newly rubbed glass tube, an electric charge—Franklin actually called it electric fire—was gradually built up on the man. This was shown by the spark seen if the man was approached by the knuckle of someone standing on the floor beyond the wax mat. Yet, Franklin noticed, it was impossible for a man to electrify himself by rubbing and then touching a glass tube that he was himself holding, even if he stood on a wax mat.[60] He suspected that this was because the rubbing does not generate new electric charge but merely shifts charge from the rubber to the object rubbed (or vice versa), so that the glass tube will transfer no more electricity to the man than it received from him in the act of rubbing. To test this hypothesis, he had two people stand on a sheet of wax; one of them rubbed a glass tube and from time to time transferred the electric charge on the glass to the other's knuckle.

He made four crucial observations:[61]

(1) After the rubbing was complete, both people on the wax appeared to be electrified to someone standing on the floor beyond the wax—i.e., a spark could be obtained by approaching either with a knuckle.

(2) If the people on the wax were touching each other during the experiment, neither subsequently appeared to be electrified.

(3) If the people on the wax touched each other after the electrification was complete, a spark passed between them that was bigger than the spark that could otherwise have been obtained between either and a person on the floor beyond the wax.

(4) After this strong spark, neither of the people on the wax appeared to be electrified when approached by the knuckle of the person on the floor.

Now consider each of these observations in the light of Franklin's hypothesis that rubbing does not create charge (electric fire) but merely shifts it between the rubber and the object rubbed, so that one ends up with an excess—*he called that being positively electrified*—and the other with a deficit—*he called that being negatively electrified.** At the start of the experiment there would be nothing to make charge flow between the three people, but after the rubbing and the transfer of charge, one of the people on the wax would have an excess and the other a deficit. These could not be obliterated by a flow of charge to or from the floor because of the insulating wax; but they could be obliterated by a flow to or from the knuckle of the person standing on the floor beyond the wax. (Any excess or deficit created in him would be rectified by the flow of charge between him and the floor.) The simultaneous existence of an excess of charge in one person and a deficit in the other explains why both people on the wax appeared to be electrified when approached by someone on the floor beyond the wax—*the first observation.* If the two people on the wax were touching during the rubbing of the glass, a flow of charge between them would prevent the excess and deficit from being built up.

* He had no way of knowing whether, when glass was rubbed by the hand, charge passed from the hand to the glass or vice versa, but he assumed that it passed from the hand to the glass; in other words that the glass was positively charged. This was an unlucky guess because we now know that, in fact, the rubbing leads to a flow of electrons—almost weightless particles bearing a uniform charge—from the glass to the hand. To fit in with the nomenclature introduced by Franklin, the charge on the electron is therefore defined as negative, and more than two and a half centuries later schoolchildren are still puzzled by the convention that an electric current in a wire is said to go from A to B if the flow of electrons is from B to A.

This explains why neither subsequently appeared electrified—*the second observation*. If the two people standing on the wax touched each other after electrification was complete, charge would have flowed from the one with an excess to the one with a deficit, thus accounting both for the large spark (*third observation*) and the subsequent lack of any evidence of electrification (*fourth observation*).

The difference between the behaviour of glass and of amber in Du Fay's experiment could be easily explained by supposing that when glass is rubbed it becomes positively charged, whereas when amber is similarly rubbed it becomes negatively charged (or vice versa). In each case the person doing the rubbing would have acquired the opposite charge.

Impressive as Franklin's theory was, it left unanswered one awkward question. If there was only one kind of electricity, wasn't it odd that an excess of electricity should have the same effect as a deficiency of electricity? Yet two positively charged bodies repel one another just like two negatively charged bodies. It was only after the discovery of current electricity at the end of the eighteenth century that Du Fay's and Franklin's ideas were reconciled by supposing, *firstly*, that uncharged bodies contain equal quantities of positive and negative electricity, which neutralize each other's properties, and *secondly*, that electrification by friction causes a transfer of charge leading to an unequal distribution of the positive and negative charges between the bodies being rubbed together. A century after Du Fay's and Franklin's work, clear evidence for the existence of positive and negative charges would come from Faraday's work on electrolysis.

So let's turn to Faraday.[62]

The son of a poor blacksmith who had moved from Westmorland to London, Faraday had minimal schooling and in 1805, at the age of 14, he became an apprentice bookbinder. Fortunately, in the bindery he was able to read widely, and was particularly influenced by Jane Marcet's *Conversations on Chemistry,* Isaac Watts' *Improvement of the Mind,* and a highly critical article on electricity in a volume of the *Encyclopaedia Britannica* that he was rebinding. He also went to lectures at the City Philosophical Society, where he met like-minded individuals, some of whom became lifelong friends. In the last year of his seven-year apprenticeship he was given tickets for four lectures to be given by Sir Humphry Davy at the Royal Institution, and he found these so attractive that he determined to give up bookbinding and try to become a scientist. He made careful notes of the four lectures but saw no way of changing career. A few months later though, Davy was temporarily blinded by an explosion in the laboratory, and was recommended Faraday as an amanuensis. He was impressed by Faraday and by the carefully bound notes of his lectures that Faraday presented to him; and in the following year, when his laboratory assistant was sacked for brawling, Davy appointed Faraday in his place.

In the autumn of that year, when Davy and his newly wed wife—a wealthy widow—were about to depart for a lengthy tour of France and Italy, Davy invited Faraday to join the party. A complication was that England and France were at war, and though Napoleon's regard for science (and for Davy) made the tour possible, Davy's passport made provision only for himself and his wife, his wife's maid and a valet. Faraday would have to go as the valet, and this led to some friction as Lady Davy tended to regard him as a servant. But on the tour he met an impressive range of European scientists, including Volta, Ampère, Gay-Lussac, Helmholtz, and Count Rumford; he studied the electric discharges of torpedo fish in Genoa, and he was present in Florence when Davy proved that a diamond consisted only of carbon, by focusing the sun's rays on it with a lens and showing that the only products were oxides of carbon.

In April 1815 Faraday, back in London, started work at the Royal Institution assisting the new Professor of Chemistry, and his work remained mostly chemical until 1821. In June of that year he married Sarah Barnard, the daughter of a silversmith whose family, like Faraday's, were active members of the Sandemanian church—a fundamentalist Christian sect, practising what they saw as an early form of Christianity.[63]

Three months later, Faraday made his first major scientific discovery. The previous year Hans Christian Oersted had shown that a compass needle could be deflected by a wire carrying an electric current. This had led to a glut of confusing articles about the phenomenon, and Faraday had been asked by the editor of the *Philosophical Magazine* to review the literature and sort out what was true. Oersted had talked about the 'electric conflict' surrounding the wire, and had said that 'this conflict performs circles'. The meaning was not clear but, mapping the distribution of magnetic force using a small compass, Faraday saw that one of the poles of the needle moved in a complete circle as the compass was moved right round the wire. It occurred to him that a single magnetic pole would rotate continuously round the wire so long as the current was flowing. You can't, of course, normally have a single magnetic pole, but you can arrange a bar magnet so that one pole is near the current-carrying wire and the other is some distance away and in line with the wire. The left-hand side of Figure 22 shows just this arrangement. Current is passed along a vertical wire dipping into a beaker of mercury; the top end of the magnet is near the wire, while the bottom end is fixed to the bottom of the beaker in such a way that it can tilt in any direction.

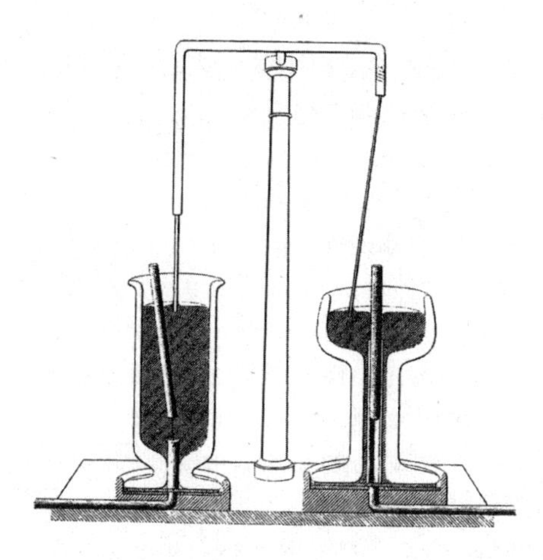

FIG 22 The first two electric motors.

Faraday found that, as he expected, so long as there was a flow of current along the wire, the top end of the magnet rotated continuously around the wire. It seemed to move along a circular 'line of force'—an expression derived from the linear arrangement of iron filings when they are scattered near a magnet. He also realized that if the electric current in the wire exerted a force on the magnet, it was conceivable that the magnet might exert a force on the current-carrying wire.** If it did, and if the magnet were fixed and the current-carrying wire were free to move, the wire should rotate continuously round the magnet. The right-hand side of Figure 22 shows the arrangement he used to confirm this prediction.

What Faraday had done in these experiments was to create the world's first two electric motors—devices that convert electrical

** Ampère's then recent finding that two parallel wires attracted each other if they were carrying currents in the same direction, but repelled each other if the currents were flowing in opposite directions, also suggested that a current-carrying wire might not only produce magnetic forces but also respond to such forces.

energy into mechanical work. No one would make electric motors like this because they would have very little power and the rotary movement would be awkward to harness, but once the principle had been established, more effective motors were soon designed by others. And Faraday is usually, and rightly, given credit for the invention.

During the next ten years Faraday was largely preoccupied with other problems and various distracting duties, but in August 1831 he turned to the problems of what we now call *electromagnetic induction*, that is, the production of a voltage across a conductor that is situated in a changing magnetic field, or is itself moving through a stationary magnetic field. Faraday was familiar with experiments in static electricity in which the redistribution of charge in an unelectrified body is caused by the approach of an electrified body, but attempts to induce a current in one wire by causing a current flow in an adjacent parallel wire had not succeeded. He had noticed, though, that the needle of the galvanometer that joined the two ends of the parallel wire gave a slight flick when the current in the first wire was switched on, and a slight flick in the opposite direction when it was switched off.

Earlier in 1831, his Dutch friend Gerritt Moll had drawn his attention to the great strength of electromagnets, and to the remarkable fact that the north and south poles of an electromagnet instantaneously exchange sites if the direction of the electric current is reversed. Faraday therefore tried the experiment illustrated in Figure 23. He took a cast-iron ring about six inches across and nearly an inch thick, and wound two separate coils of insulated copper wire round opposite halves of the ring. The coil on the left (what we should now call the primary coil) could be connected through a switch to a powerful battery; that on the right (the secondary coil) was connected to a galvanometer. When a battery consisting of 'one hundred pair of plates' was

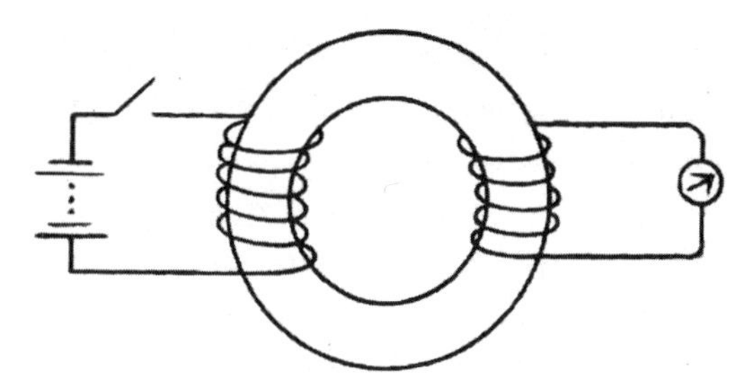

FIG 23 The first induction coil. The point of this arrangement is that the magnetic field generated by the current passing through the left-hand coil is almost entirely contained within the iron ring and so passes through the right-hand coil. This occurs because the individual iron atoms behave like tiny magnets and arrange themselves in circles centred on the midpoint of the ring.

connected to the primary coil 'the impulse at the galvanometer...was so great as to make the galvanometer needle spin round rapidly four or five times before the air and terrestrial magnetism could reduce its motion to a mere oscillation.' While a steady current was flowing, the needle remained stationary, but when the battery was disconnected, the needle span several times in the opposite direction. Clearly, it was the *changes* in the magnetic field in the region of the secondary coil that induced the transient currents recorded by the galvanometer.

By having more turns in the secondary coil than in the primary, it was possible to produce higher voltages than those in the primary coil, and within a few years of Faraday's experiment, others would be using 'induction coils' to generate brief high-voltage shocks for various purposes. Much later, when the use of alternating currents had become widespread, induction coils—by then more often called 'transformers'—were used both to increase the voltage to aid transmission over long distances, and to reduce it again to a safe level for domestic use. Nowadays, we find that our houses tend to be full of small transformers, reducing the voltage still further for telephones, toothbrushes, children's toys, laptop computers, scanners, printers, Christmas lights, burglar alarms, doorbells, or cameras.

In the experiment shown in Figure 23, an electric current in one conductor was used to change the magnetic field in the region of another conductor, so inducing a transient electric current in the second conductor. But Faraday went on to show that an electric current could be induced more simply just by moving a loop of wire between the poles of a magnet, or pushing a bar magnet into the end of a coil of wire. He showed that the voltage produced was proportional to the rate of change of the magnetic field, and that the direction of the current was determined by the direction of the movement. If a bar magnet was inserted into one end of a coil of wire connected to a galvanometer, the needle moved in one direction during the insertion, and then returned to its original position; if the magnet was now withdrawn, the needle moved in the opposite direction during the withdrawal, and again resumed its original position.

As well as inventing electric motors and induction coils, Faraday also invented the dynamo—a device that uses mechanical work to generate electricity. Since most modern dynamos involve coils of wire rotating in magnetic fields, their invention might seem a likely extension of the experiments that I have been describing; and in fact in 1832, a Frenchman called Hippolyte Pixii, inspired by Faraday's experiments, arranged to rotate a horseshoe magnet round a vertical axis so that its two poles rotated in a circle just below two connected vertical coils of copper wire. The result was an alternating current in the wire.[64] This would have been the world's first dynamo but Pixii had been preceded by Faraday using a quite different and rather surprising approach.

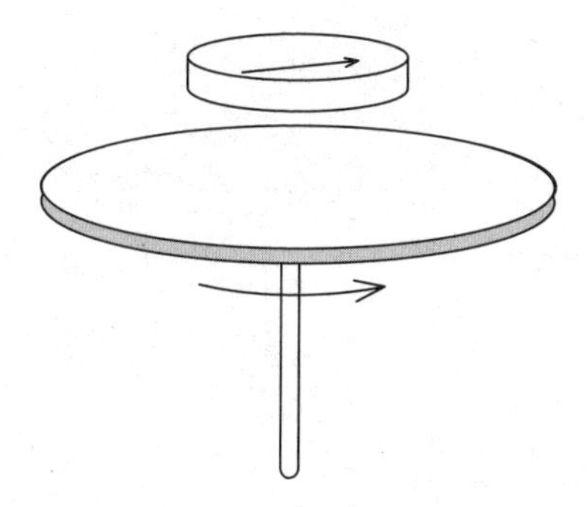

FIG 24 'Arago's wheel'.

In 1824, Francois Arago in Paris described a phenomenon that became known as Arago's wheel. He showed that when a horizontal copper disc was rotated around a vertical axis, the needle of a compass mounted just above the disc rotated in the same direction as the disc (see Figure 24).

Since copper is not magnetic, and there was not the slightest attraction or repulsion between the *stationary* copper disc and the compass needle, this was puzzling, and no satisfactory explanation was available. In October 1831, though, Faraday realized that a disc rotating just below a compass needle was a conductor, each part of which was moving through a magnetic field. In the light of his recent experiments this meant that an electric current must be generated in the disc; such a current would in turn create a magnetic field, and this field would exert a force on the needle. This provided a neat explanation of Arago's wheel, but it did more than that. If instead of rotating the disc below a compass needle, he rotated it between the poles of a powerful magnet, as in Figure 25, he should, he thought, be able to produce a much stronger current.

He tried the experiment, using sliding contacts to connect a galvanometer to different pairs of points on the rim of the disc,

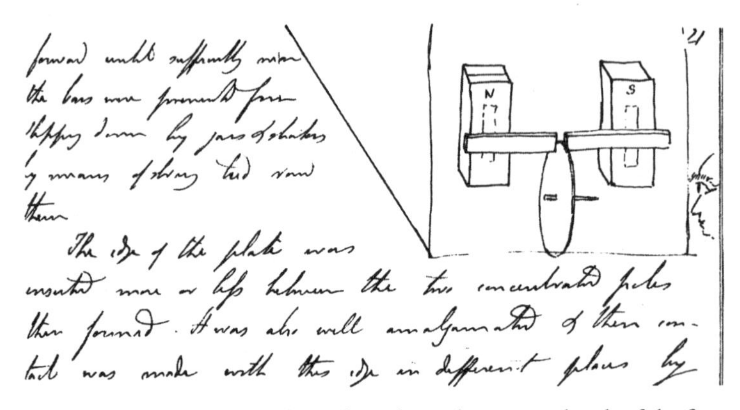

forward until sufficiently near the bars were prevented from slipping down by pairs of [illegible] by means of string tied round them

The edge of the plate was inserted more or less between the two concentrated poles then formed. It was also well amalgamated & then contact was made with this edge in different places by

FIG 25 The top of a page of Faraday's diary showing a sketch of the first dynamo.

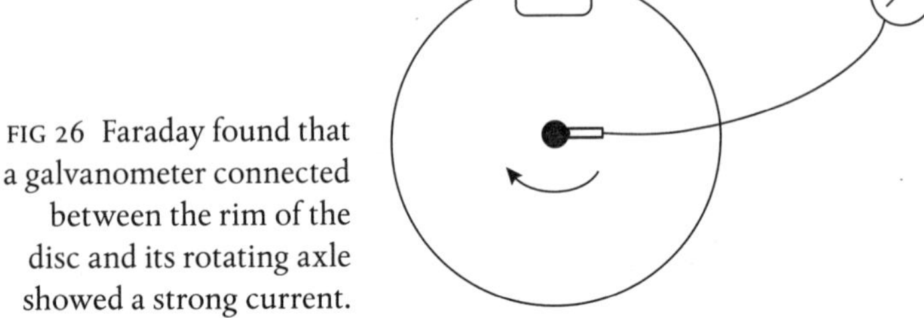

FIG 26 Faraday found that a galvanometer connected between the rim of the disc and its rotating axle showed a strong current.

but obtained only feeble results. However, when he placed one wire from the galvanometer in sliding contact with the axle on which the disc turned and the other wire in sliding contact with the disc's rim (Figure 26), he detected a steady current.[65] He had invented the first dynamo.

Interestingly, whether in creating the electric motor, the transformer, or the dynamo, Faraday was more concerned with understanding what was going on than with looking for ways of improving the machine's performance. In his Bakerian lecture to the Royal Society, given in January 1832, he does discuss briefly a possible improvement in his dynamo—he suggests multiple parallel discs with alternate discs moving in opposite directions—but he goes on:

> I have rather, however, been desirous of discovering new facts and new relations dependent on magneto-electric induction, than of exalting the force of those already obtained; being assured that the latter would find their full development hereafter.[66]

His confidence in the hereafter was not misplaced. And he was not entirely uninterested in possible applications. There is a story that when Sir Robert Peel visited the Royal Institution, he pointed at Faraday's dynamo asking what use it was. 'I know not,' Faraday is supposed to have said, 'but I wager that one day your government will tax it.' In fact, it was not until long after Peel and Faraday were both dead that the electrical power industry became worth taxing.[67]

By the end of 1832, Faraday had transferred his attention from electromagnetic induction to what we now call electrolysis—the breakdown of a substance resulting from the passage of an electric current through it (excluding breakdown caused by heat generated by the current). In 1800, shortly after Volta had published an account of his pile,† two Englishmen, William Nicholson and Anthony Carlisle,[68] made a similar pile and showed that if two platinum or gold wires connected to the two ends of the pile were immersed at some distance apart in a tube of drinking water,[69] bubbles of hydrogen were formed on one wire

† See footnote in Chapter 4, page 77.

and bubbles of oxygen on the other. The volume of hydrogen produced was about double the volume of oxygen, and since that is the ratio of the amounts of hydrogen and oxygen (expressed as volumes at a given pressure) needed to make water, it seemed that the current was decomposing the water into its constituent elements. What was puzzling, though, was that the hydrogen was liberated at the surface of one wire and the oxygen at the surface of the other. If the current tore some of the water molecules apart, why didn't the two gases appear together? How could the two gases travel in opposite directions towards the two wires without leaving any trace until they were close to the wire? At the beginning of 1833 this problem was still unsolved.

In that year Faraday began a serious quantitative study of electrolysis. He used a battery of voltaic cells to pass currents between metal electrodes immersed in solutions of acids, alkalis, or salts, and he was interested in the relation between the amount of current passing through the solution and the amount of chemical change that that current produced. A particularly elegant experiment is illustrated in Figure 27. Using a branched circuit, he passed current generated by a battery through two identical electrolytic cells (A and B). (An electrolytic cell is simply a vessel containing a solution through which an electric current is passed between two electrodes, resulting in the partial breakdown of the dissolved substance.) The reunited current was then passed through a third identical electrolytic cell (C), and back to the other pole of the battery. With this arrangement, in a given time, the total amount of electricity passing through the third cell (C) must be the sum of the amounts passing through each of the cells A and B. And from the symmetry of the situation, the currents passing through cells A and B will be equal. What Faraday found was that the amount of each of the substances liberated in cell C was just equal to the sum of the (equal) amounts liberated in the cells A and B.

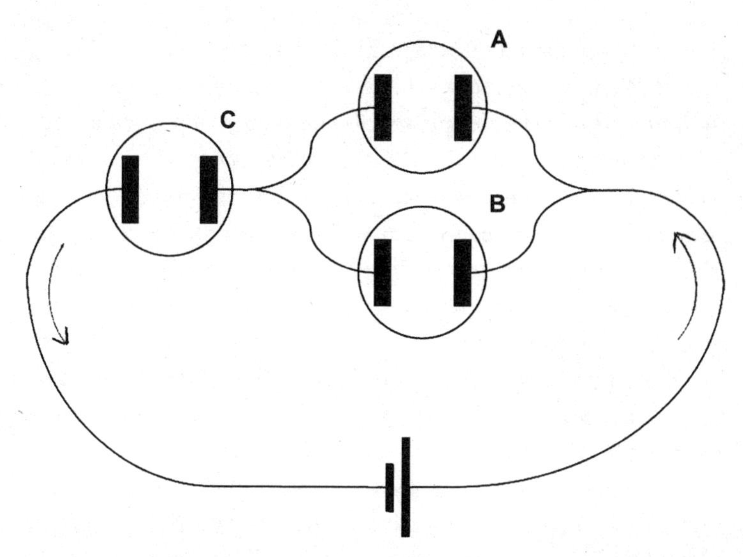

FIG 27 Faraday's experiment to investigate the relation between the chemical change occurring in an electrolytic cell and the total charge passed through the cell.

In further experiments he found that, for a given current, the amount of chemical change was proportional to the duration of the current and was totally unaffected by the size or shape of the electrodes. Clearly *for any given cell* the amount of chemical change was proportional to the amount of electric charge passed through the cell—a finding that became known as *Faraday's first law of electrolysis*.

This finding prompted the question: if you pass the same amount of charge through electrolytic cells containing different solutions, what determines the amount of chemical change that the passage of charge causes in each? Faraday therefore did another series of experiments—as informative as they were simple—in which he passed the *same* current for the *same* time through a number of different electrolytic cells connected in series, and compared the amounts of chemical change occurring in each. What he found was that, for a given passage of charge, the amounts of different substances liberated or deposited in each cell were proportional to their 'chemical equivalent weights'. This became known as *Faraday's second law of electrolysis*, but since the concept of chemical equivalent weights seems to be ignored in modern school chemistry textbooks, it needs some explanation.

The idea that when substances react together they do so only in definite proportions by weight goes back to Al-Jildaki in fourteenth-century Egypt, but the principle was rediscovered by Henry Cavendish in 1766, and was well established by the end of the eighteenth century. It was convenient to choose a standard weight of one particular substance in terms of which the 'equivalent weight' of other substances could be expressed, and the standard first chosen was 1 unit (by weight) of hydrogen. Water contains 8 parts by weight of oxygen for each part by weight of hydrogen, so the equivalent weight of oxygen was 8, and—since water contains no other elements—the equivalent weight of water was 9. Of course, not all substances combine with hydrogen or liberate hydrogen from its compounds, but their equivalent weights can be calculated indirectly. For example, copper neither combines with hydrogen nor liberates hydrogen from acids, but it does form an oxide. In copper oxide, 31.8 grams of copper are combined with 8 grams of oxygen, so the chemical equivalent weight of copper is taken to be 31.8. Arguing in this way, Faraday's slightly older Swedish contemporary J.J. Berzelius and his pupils were able to create a general table of equivalent weights. Now back to Faraday.

The fact that the same chemical compound always consists of the same elements combined together in the same proportions by weight was, of course, strong evidence that the chemical elements consisted of atoms—that is, particles that are indivisible (or if divided no longer possess the properties of the original substance) and that have a particular weight. Faraday's finding that, for a given passage of charge, the amounts of the different substances liberated in the individual electrolytic cells were *in proportion to the substances' equivalent weights*, showed that the equivalent weights of different charged particles must carry similar amounts of charge.

Faraday assumed that when electricity was passed through a dilute solution of, say, hydrochloric acid, some of the molecules of the acid (which are electrically neutral) separated into positively charged hydrogen ions (H^+) and negatively charged chloride ions (Cl^-). (The word *ion*—related to the Greek word for wanderer—was suggested to him by William Whewell, the omniscient Master of Trinity College, Cambridge.) Because of their charge, the hydrogen ions throughout the solution would tend to migrate towards the wire (*electrode* was the word suggested by Whewell) connected to the negative pole of the battery, and the chloride ions would tend to migrate towards the electrode connected to the positive pole. In the bulk of the solution, the situation that each hydrogen ion and each chloride ion found itself in would not change significantly—each ion would, as it were, continuously change partners—but at the electrodes things would happen. At the negative electrode, hydrogen ions would lose their positive charge and become hydrogen gas. At the positive electrode, chloride ions would lose their negative charge and become chlorine gas—assuming that the electrode was made of a material that is not attacked by chlorine.

An explanation of this kind can account for the finding that in the electrolysis of very dilute sulphuric acid, hydrogen gas appears only at the negative electrode and not in the bulk of the solution. But the electrolysis of very dilute sulphuric acid also leads to the release of oxygen at the positive electrode. Why does oxygen appear at all, and why does it only appear at the positive electrode? The answer—we now know—is that in a solution of very dilute sulphuric acid there will be many negatively charged sulphate ions (SO_4^{2-}) and a few negatively charge hydroxyl ions (OH^-)—formed by the dissociation of water into H^+ and OH^- ions. In the bulk of the solution, the great majority of the negatively charged ions tending to migrate towards the positive electrode will be sulphate ions, but at the surface of the electrode it is apparently much easier for hydroxyl ions to give up their negative charge than it is for sulphate ions. The result is that although they are in a small minority it is largely the hydroxyl ions that give up their charge; when four hydroxyl ions have lost their charge they can form two molecules of water and one molecule of oxygen gas (O_2).

Faraday did not have this detailed information; but at an early stage he realized that, though in the bulk of the solution the current would be carried by the available ions, what happened at the electrodes would depend not only on which ions were arriving or leaving but also on secondary reactions involving the material of the electrodes. He knew, for example, that Carlisle and Nicholson had found that if an electric current was passed through very dilute sulphuric acid using brass wires instead of platinum wires, hydrogen gas was released at the negative wire; while at the positive wire no oxygen gas was released, but the brass lost its sheen and was darkened, as if the metal had been oxidized by the newly formed oxygen atoms.

Like his pioneering work on electromagnetism that led to electric motors, transformers, and dynamos, Faraday's work on electrolysis had far-reaching practical consequences. He showed, for example, that if an electric current was passed between two metal electrodes immersed in a solution of copper sulphate, copper was deposited on the negative electrode. If the positive electrode was made of copper, at its surface copper atoms would pass into solution as positively charged copper ions, and the overall effect would be a transfer of copper through the cell from the positive to the negative electrode. Similar transfers could be arranged using suitable salts of silver or gold, and within ten years electroplating factories had been established in England, France, and Russia. Birmingham became the leading centre for 'electrogilding and silvering', replacing Sheffield, which had long been famous for the production of what we now call Old Sheffield Plate. That had been made by a traditional (and extravagant) method in which a thin sheet of silver was fused onto a thicker sheet of mixed copper and brass, and then repeatedly squeezed between rollers with occasional heating to soften the metal and remove internal stresses.[70]

The 1830s had been a particularly fruitful period in Faraday's life, but late in 1839 he had a sudden and severe attack of vertigo, and though he recovered from this initial attack, his health was never fully restored. He gave up many of his duties at the Royal Institution, and although during the next five years he invented a new kind of chimney for lighthouses, which prevented smoke from contaminating the lantern, and he undertook an official investigation into the cause of an explosion at the Royal Gunpowder Mills in Waltham Abbey,[71] it looked as though his scientific career was largely over. In 1845, though, he did an experiment as significant as any in his career. He used a powerful electromagnet to apply a strong magnetic field to a beam of polarized light passing through a glass block. He found that the plane of polarization of the light was rotated, and that the angle of rotation was proportional to the strength of the field.[72] To appreciate the importance of this dramatic link between magnetism and light, it is necessary to know more about light, so it is time to move to the next chapter.[73]

6

THROWING LIGHT ON LIGHT: WITH THE STORY OF THOMAS YOUNG

'We all *know* what light is; but it is not easy to *tell* what it is.'
Samuel Johnson, 1776

Thomas Young, wrote Isaac Asimov, 'was the best kind of infant prodigy, the kind that matures into an adult prodigy.'[74] That is true, but what makes Young so extraordinary is the variety of the problems for which he produced elegant solutions—ranging from the nature of light and the mechanism by which we recognize colours, to deciphering the Rosetta stone and the nature of ancient Egyptian hieroglyphics.

Born in the village of Milverton, in Somerset, in 1773, Thomas was the first of ten children of a Quaker mercer and country banker.[75] Soon after his birth he was taken to the Minehead house of his maternal grandfather, also a Quaker merchant, where he spent most of his early childhood. He later wrote warmly of this grandfather, who was interested in his education, and his progress as an infant seems almost unbelievable. He tells us that he 'had learnt to read pretty fluently by the time he was 2', and he is supposed to have recited nearly all 430 lines of Goldsmith's poem

The Deserted Village at the age of 6. His early schooling was not happy, but when he was nearly 9 he was transferred to a school in Compton, Dorsetshire, where again he flourished. Between the ages of 9 and 13, he studied the Greek and Latin classics, elementary mathematics, French, and some Italian and Hebrew. Whether there was any formal teaching of science is not clear, but the school usher, Josiah Jeffrey, lent him Benjamin Martin's *Lectures on Natural Philosophy*, which introduced him to Newton and whose section on optics fascinated him—though he found Martin's introduction to the method of fluxions (Newton's term for calculus) incomprehensible. Jeffrey also taught him how to draw, mix colours, bind books, use a lathe, grind lenses, and make telescopes.

When he was at home during a school holiday, his father lent him Joseph Priestley's book on air, starting his interest in chemistry; and at Minehead a 'philosophical and meteorological saddler' lent him a quadrant, which he used to measure the height of the local hills. In the autobiographical sketch that he wrote two or three years before he died, he tells us (writing in the third person, and originally in Latin) that:

> He had imbibed from a conversation with [a Minehead resident] a fancy for studying botany; and for the sake of examining plants, he wished to make a microscope from the descriptions of Benjamin Martin; the botany was then forgotten for the sake of optics; and in order to make his microscope he thought it necessary to have a lathe; the lathe was made, and optics were then thrown aside in favour of turning;...

In 1786, aged 13, he finished school and returning home settled down to improving his Hebrew—he read the first thirty chapters of Genesis—and to developing his skills in using the lathe and in making telescopes. Prompted by a discussion at dinner about whether eastern languages differed from one another as much as

European languages, he began to learn Arabic and Persian, he borrowed grammars of Chaldee, Syriac, and Samaritan, and he derived 'extraordinary pleasure'—the words are his own—from a book that gave the 'Lord's Prayer in more than 100 languages'.

In the following year his life again changed dramatically, as a result of Quaker networking. David Barclay, of the Quaker banking and brewing family, selected Thomas Young as a suitable companion for his grandson, Hudson Gurney, who was to be taught by a private tutor while living in Barclay's country house. The arrangement was a total success. The tutor, John Hodgkin (later the father of the Hodgkin of Hodgkin's disease), was a young classical scholar who taught the boys 'the art of writing English', and encouraged Thomas' writing of Greek script so successfully that the book of ancient Greek texts that Hodgkin produced was illustrated with examples written by Thomas. This detailed attention to an ancient script could hardly have been more appropriate for a boy who, as a man, would help decipher the Rosetta stone, with its inscription in three scripts—ancient Egyptian hieroglyphics, later demotic Egyptian, and ancient Greek.[76] But Thomas did not neglect the sciences. He studied botany and zoology—particularly entomology—as well as mathematics and physics; and, astonishingly, at the age of 17 he read and understood[77] Newton's *Principia* and *Opticks*.

Nor was he just a polymath. As a good Quaker, from the age of 14 he refused to eat sugar from the West Indian plantations, which was 'the produce of the labour of slaves'. David Barclay shared Thomas' aversion to slavery, spending about £3000 liberating thirty slaves in a Jamaican property of which he had become owner, and chartering a vessel to transfer them to Philadelphia where they were 'put out by Friends as apprentices on equitable terms.'[78]

When he was about 15, Thomas developed consumption. He was treated by Baron Thomas Dimsdale—the barony had been a reward from Catherine the Great for inoculating her against smallpox—and by Richard Brocklesby, his own great-uncle, who was a fashionable physician in London with Edmund Burke and Samuel Johnson among his patients. Over the next two years Thomas recovered completely, and the main effect of his illness was that Brocklesby became very interested in his accomplishments[79] and concerned about his future. Feeling that the right career for his great-nephew was medicine, and having no children of his own, Brocklesby offered to pay for Thomas' further education and to leave him part of his estate. In the autumn of 1792, aged 19, Thomas moved to London, took lodgings in Westminster, and began life as a medical student.

He started at the School of Anatomy founded by William Hunter, and in his second year he also enrolled at St Bartholomew's Hospital, the oldest of the London Hospitals and at that time the only one that arranged a regular course of lectures for medical students. Dissecting an ox eye at the Hunterian School, he convinced himself that the lens of the eye was *itself* muscular; it could therefore be responsible for the change of shape that Descartes had suggested enables the eye to focus on objects at different distances. In May 1793, still only 20, he reported his findings to the Royal Society, which was somewhat embarrassing as his hypothesis turned out to be neither new nor true. Nevertheless, he was elected to the Society the following year, proposed by a group of Fellows including his great-uncle and a number of other distinguished physicians. (It was not until the later part of the nineteenth century that a distinguished record in research became almost essential for election to the Royal Society.)

After two years in London, Young spent a year in Edinburgh, whose medical school was then at the height of its fame, followed by a year at Göttingen—both universities, unlike Oxford and Cambridge, open to Quakers. At Edinburgh he continued to work hard at both medical and more general subjects, but he also became interested in the theatre—he saw Mrs Siddons in seven different plays—he learnt to play the flute, and to dance, and he gave up Quaker dress. The strict Quaker was weakening. At Göttingen he continued to work on a very broad front—medical, cultural, literary—and he added playing the clavichord and drawing to his hobbies. To get his doctorate, he needed to sit an examination lasting four to five hours, on 'practical physic, surgery, anatomy, chymistry, materia medica, and physiology', but, he tells us, 'the four examiners were seated round a table, well furnished with cakes, sweetmeats, and wine, which helped to pass the time agreeably.'[80] In addition, he submitted an eighty-page dissertation *On the Preservative Powers of the Human Body*,[81] and gave a public lecture in which he talked about the formation of the human voice. He suggested that an alphabet of 47 symbols, used alone or in combination, should be able to express every sound that the organs of the human voice are capable of making. Such an alphabet, he felt, would be particularly useful for recording languages previously unwritten, for example those met by missionaries. Again we see activity appropriate for the man who would later begin to decipher the Rosetta stone; but that is just one connection. Young himself tells us that it was his interest in the formation of the human voice that led him to investigate the nature of sound and the laws of its propagation; which in turn led him to consider analogous propositions about the nature of light.[82]

Having achieved his doctorate in 'Physic, Surgery and Man-Midwifery' from Göttingen in 1796, Young had intended to travel widely in Europe, and then return to London to begin life as a physician. But the French revolutionary wars were splitting Europe—indeed 1796 was the year in which Count Rumford was summoned back to Munich to help protect it against the French and the Austrians—so Young had to restrict his travels to visiting German towns. When he did return to England, early in the following year, he faced an unexpected personal problem; new rules from the College of Physicians meant that he was not allowed to practise as a Licentiate in London until he had spent two consecutive years at the same University.[83] Brocklesby was keen that the university should be Cambridge, and the idea had obvious attractions, but Quakers of course could not be admitted. So—like Henri IV reckoning that 'Paris was worth a Mass'—Young finally cut his links with the Quakers and joined the Church of England. In March 1797, he became a Fellow-commoner—a student but with many of the privileges of a Fellow—at Emmanuel College, Cambridge; its Master was an old friend of Brocklesby and had been one of the Fellows of the Royal Society who had proposed Young for election. During the next two years, with little medical teaching available, Young spent most of his working time reading, writing, thinking, and doing occasional experiments to explore the nature of sound and light. He also led an active social life, finding friends among the Masters and Fellows of several Colleges; in his own College he was elected President of the Fellows' Parlour, and his name appears from time to time in the Fellows' betting book. Among undergraduates he was known as 'phenomenon Young'.

By the end of 1799, Young had completed his two years in Cambridge and returned to London. In 1797, Brocklesby had died suddenly, leaving Young his London house and its furniture, his library, a collection of pictures and £10,000 pounds—a sound basis for beginning a medical career. But though Young had his doctorate from Göttingen, the rules of the University of Cambridge meant that he would have to wait until 1803 before he acquired a bachelor's degree in medicine, and until 1808 before becoming a Cambridge MD. He would be 36 before he was elected a Fellow of the Royal College of Physicians. And, despite Brocklesby's generosity, even after selling Brocklesby's fashionable house near Park Lane and buying a more modest house in the less fashionable Welbeck Street, he needed an income. He started to practise medicine in 1800, but progress was slow, and he had so much leisure to attend meetings of the Royal Society that he became well known to Joseph Banks—then President—and Count Rumford. When, in 1801, he was offered a Professorship in Natural Philosophy with a salary of £300 per annum, at Rumford's year-old Royal Institution, he was happy to accept it.

The job involved giving fifty to sixty popular lectures on 'natural philosophy and the mechanical arts' to the Institution's members. This needed a great deal of work, and when we now read the published version of those lectures they are enormously impressive in both their range and their content.[84] But they were quite unsuitable for a lay audience, and after holding his professorship for two years Young resigned, probably under pressure from the managers of the Institution, and returned to his sluggish medical practice. The following year, 1804, he was appointed Foreign Secretary of the Royal Society, and in that same year he married Eliza Maxwell, a young Scotswoman of aristocratic descent. Both the appointment and the marriage were highly successful, though the marriage produced no children. What is remarkable is that the period between 1797 and 1804, with all its uncertainties and stresses, is just the period when Young was doing what we now recognize as his most important and most elegant scientific work—proving the undulatory theory of light, and suggesting the trichromatic theory of colour vision. His thoughts on colour vision will be discussed in a later chapter, but it is time to look at his work on the undulatory theory.

Towards the end of the eighteenth century there were still two main theories about the nature of light, both of them with a long history and neither of them wholly satisfactory. The first—the so-called *corpuscular* theory—held that a ray of light consisted of a stream of particles; the second—the *undulatory* theory—that it consisted of a succession of waves, something like sound waves.* But since it was then thought that all waves needed a medium through which they could be transmitted, and (unlike sound) light appeared to be transmitted through what was usually regarded as empty space, to account for the transmission of light waves it seemed necessary to postulate the existence of some kind of all-pervading ether. In the seventeenth century, the corpuscular theory had been much favoured by Newton, and the undulatory theory by the Dutch physicist Christiaan Huygens, though it was clear that some features of light were more easily explained by one theory and other features by the other.

A striking feature of light, obvious from the appearance of shafts of sunlight breaking though the clouds, or from the sharpness of shadows, or from our inability to see round corners, is that it travels in straight lines. That is to be expected of streams of weightless particles, but it is not what we are familiar with in dealing with waves. Both water waves and sound waves spread round corners. When waves strike the hull of a boat they sweep

* Sound waves are the waves of alternate compression and decompression of the air (or other medium) that spread in all directions from the vibrating string, or vocal cords, or tuning-fork, or dropped kettle-lid, or whatever it is that is the source of the sound.

round it; we hear the church bells when buildings lie between us and the belfry; we can hear people in the next room through a slightly open door even if they are totally out of sight. This is why Newton was so reluctant to accept the undulatory theory for light; and he remained reluctant even though he confirmed the finding of the Italian physicist Francesco Grimaldi that the shadow cast by a beam of light passing the edge of a knife or a needle was not absolutely sharp and there were coloured fringes running parallel to the edge of the shadow—a phenomenon that Grimaldi called *diffraction* (from the Latin *diffringere,* to shatter) and which suggested an undulatory phenomenon.

What wasn't realized in the seventeenth century was that the reason light seems to travel in straight lines whereas water waves and sounds bend round corners is *not* that water waves and sound are waves whereas light is not, but that there is a great difference in wavelengths. The two photographs in Figure 28 show what happens when straight water waves in a ripple tank meet a barrier with a gap. In the left-hand photograph, the gap is less than two wavelengths wide, and beyond the gap the waves spread out in semicircles, centred on the opening as if it were a point source of the waves. In the right-hand photograph, the gap is nearly twenty times the wavelength, and the waves passing through the gap continue in a straight line with no change of direction. Only at the edges is there a little outward spread. In other words, diffraction is prominent when the gap is relatively small, and marginal when it is relatively very large. Now audible sound waves vary in wavelength from about 17 mm, for the highest pitches we can hear, to about 17 metres for the lowest. So (assuming for the moment that in this respect sound waves behave like water waves) it is not surprising that we can hear round corners. In contrast, the wavelengths of visible light waves vary from about 0.4 micrometres at the violet end of the spectrum to about

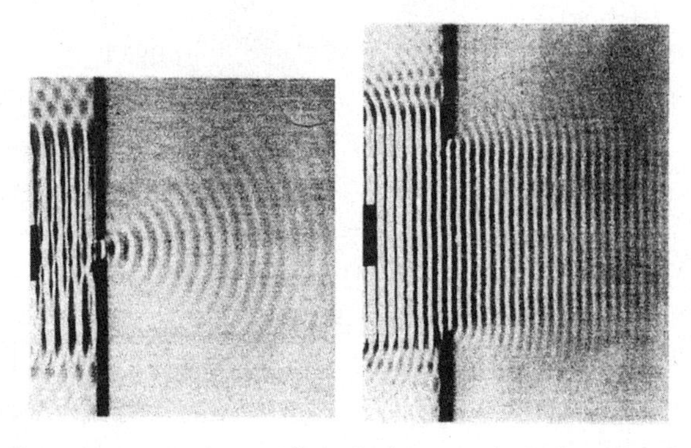

FIG 28 Experiments with a ripple tank showing the behaviour of straight waves when they meet a barrier with narrow or wide openings.

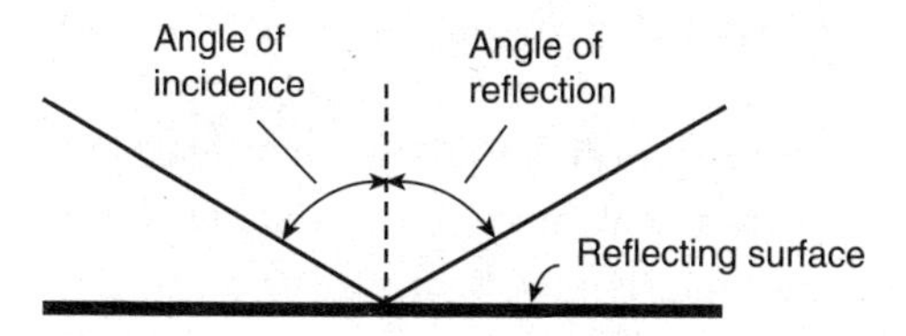

FIG 29 The behaviour of a reflected beam of light.

0.7 micrometres at the red end. So (again assuming that light waves behave like water waves) it is not surprising that in everyday life light nearly always appears to travel in straight lines.

A second feature of light is that when a ray of light is reflected from a flat mirror, the 'angle of incidence' and the 'angle of reflection' are always equal (see Figure 29). This feature was easily explained by either theory. In the corpuscular theory, the particles simply bounce like billiard balls;[85] by the undulatory theory, the waves are reflected just as water waves are reflected when they strike a flat surface.

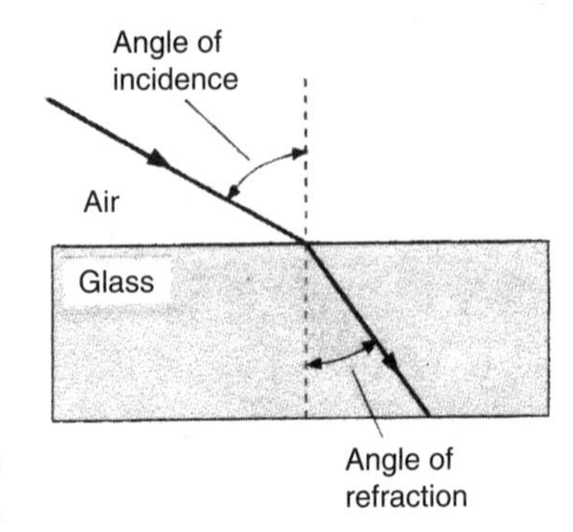

FIG 30 The behaviour of a refracted beam of light.

A third feature, recognized by Ptolemy in the second century BC, is that when a ray of light strikes the flat surface of a transparent and more dense medium, such as glass or water, the direction of the ray of light is changed as it enters the denser medium, and the 'angle of refraction' is less than the angle of incidence (see Figure 30). Huygens explained this on the undulatory theory by the plausible hypothesis that light travelled more slowly in the more dense medium. He suggested that the spread of a wave occurs because *each point on the wavefront at any moment acts as a point source of new circular waves,* and the tangent to these new circular waves defines the new wavefront (see Figure 31, in which the new circular waves are referred to as secondary wavelets). If, when the ray of light enters the denser medium, the waves spread more slowly, the ray will be refracted (see Figure 32).

It was more difficult to explain refraction with the corpuscular theory. Newton suggested that when the corpuscles enter the denser medium they are acted on by a force pulling them towards the normal (i.e., the line perpendicular to the surface) and increasing their velocity. There was, though, no independent evidence for such a force, and it seemed unlikely that a stream of particles would move faster in a more dense medium.

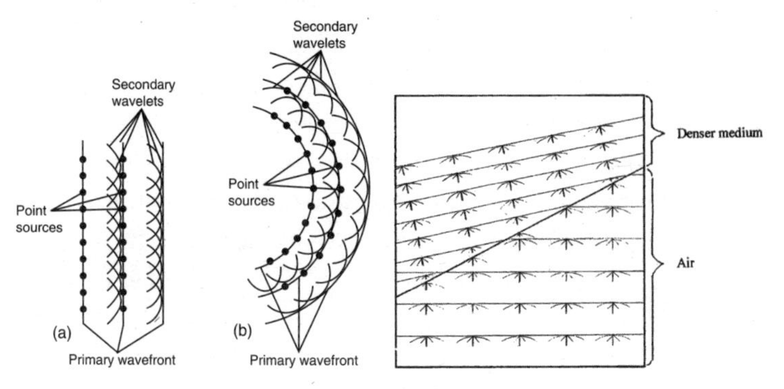

FIG 31 Huygens' construction explaining the propagation of a primary wavefront.

FIG 32 A diagram, based on Huygens' construction, showing how the slowing of light waves as they obliquely enter a denser medium changes the direction of the wavefronts.

Newton had even more difficulty in explaining a fourth feature of light. When the sun shines on water some is refracted and some is reflected. With water waves and sound waves such split behaviour is familiar, but why should some corpuscles behave one way and some another way? Newton suggested that the corpuscles could exist in different states. When it struck the water, a corpuscle that was in a 'Fit of easy Reflection' would be reflected, while a corpuscle that was in a 'Fit of easy Transmission' would be refracted. Curiously, he even suggested that 'Waves of Vibrations, or Tremors' of some kind might determine which of the two states any corpuscle was in at any moment, but his two-state hypothesis still looked very contrived.[86]

Huygens died in 1695, and Newton in 1727, yet the confusion about the nature of light continued until the work of Thomas Young at the beginning of the nineteenth century.

The crux of Young's approach was to consider what happens when two series of waves cross one another. According to an

Emmanuel College legend, Young first became interested in what later were called interference patterns by watching the ripples generated by a pair of swans in an oblong pond in the College garden. What is certainly true is that in his Royal Institution lectures he used a ripple tank to demonstrate such patterns, and that he liked arguing from observations on water waves and sound waves to what might be expected of the hypothetical light waves. Of course, water waves are 'transverse waves'—that is, the water molecules in a wave move (alternately up and down) in directions at right angles to the direction of movement of the wave, whereas sound waves are longitudinal waves, that is, the predominant movement of the air molecules is alternately forwards (in the direction that the wave is moving) or backwards (in the reverse direction)—see Figure 33. This makes sense if you think of what happens when you strike the prong of a tuning fork or pluck the string of a violin. The vibration alternately compresses and decompresses the air in contact with the vibrating surface and a succession of compression waves move outward in concentric spheres. When these waves reach our ears we hear a note, but what has moved from the tuning fork or violin string to our eardrums is merely a succession of local disturbances in the air molecules. As with water waves, there is no transfer of anything material.

Although the movement of air particles in sound waves is not transverse, Young found it helpful to think of the forwards and backwards movements at a particular point in the path of the wave as if they were plotted on the (vertical) *y*-axis of a graph, with time represented on the (horizontal) *x*-axis. The resulting graph of the sound wave would then resemble the graph of a water wave: in both cases the graph would show the way in which, at a particular point, the deviation of particles from their resting position changes with time.

(a) **Longitudinal waves, such as sound waves in air, or compression waves in a slinky**

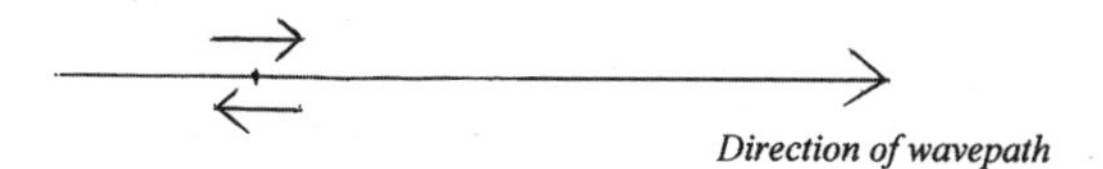

The arrows show the small forward and backward vibratory movements of molecules, at a given point on the wavepath

(b) **Transverse waves in water**

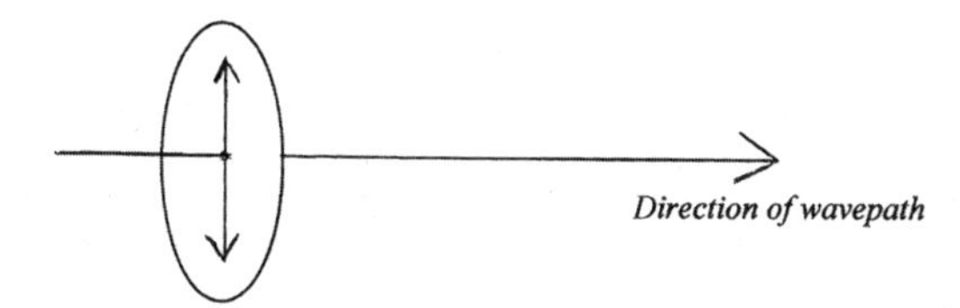

The arrows show the movement of water molecules at a given point in the wavepath. They move alternately up and down, in a plane perpendicular to the wavepath

(c) **Transverse waves in a rope shaken by hand**

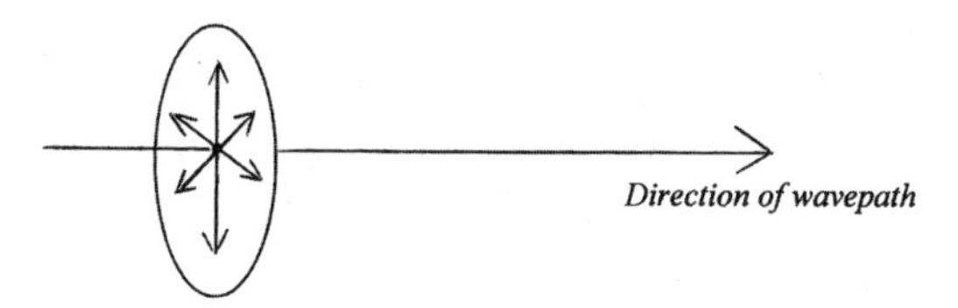

Depending on the shaker, the alternate movements of the rope particles at any given point in the wavepath can be in any direction within a plane perpendicular to that wavepath

FIG 33 Longitudinal and transverse waves.

An early product of Young's approach was his elegant explanation of the beats that are heard when two steady notes of similar loudness and nearly the same pitch are played together—the beats that are so familiar to piano tuners.[87] He argued that when sound waves from two different sources cross in space, each air particle in the path of both waves must take part in both motions.** If the two notes being played are *identical* in pitch, the addition of the two motions will give a steady note of the same pitch. But if there is a *slight difference* between the two pitches—and, therefore, a slight difference between the frequencies (and hence the wavelengths†) of the two waves—the relative timing of the peaks and troughs in the two waves will change with time. When the peaks occur together and the troughs occur together, he argued, the motions will be additive and the note will be louder. When the peaks in one wave accompany the troughs in the other, and vice versa, the motions will tend to cancel out and the note will be quieter. And this cycle of events will occur repeatedly. The result of all this, he pointed out, is that the note heard will be of intermediate pitch and will fluctuate in intensity, with a louder beat in each cycle. The closer the notes are in pitch, the longer the cycle will take, and the longer the intervals will be between the beats. Figure 34 shows the calculated result of adding two sound waves, identical in amplitude and form but differing slightly in pitch—and therefore in wavelength. The top graph shows the waveforms of the two individual waves; the bottom graph shows the waveform of their combined effect.

** This follows from the *principle of superposition of waves,* which states that when two (or more) waves overlap, the resultant wave is the algebraic sum of the individual waves.

† The speed of a wave is given by the product of the frequency and the wavelength. Since the speed of a sound wave is constant in a given medium, the frequency and the wavelength are inversely related. One tends to think of frequency rather than wavelength when talking about sound, because the pitch becomes higher if the frequency is increased, but falls if the wavelength is increased.

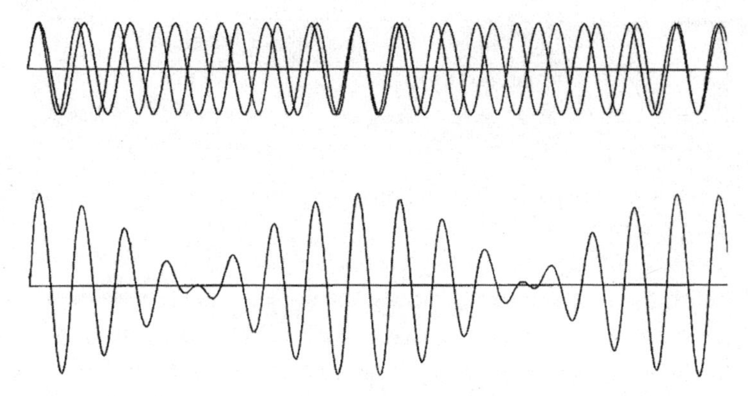

FIG 34 The addition of two similar waves with slightly different frequencies to give 'beats'.

Young had no way of measuring the movements of air associated with the sound waves, but nowadays we can record the vibrations of the diaphragm of a microphone exposed to the sound by displaying the microphone's output on an oscilloscope. The results of experiments in which the notes of similar but not identical pitch are recorded, and the waveforms of the individual sounds and of the combined sound are displayed, show patterns very like the pattern shown in Figure 34.

Note that Young's elegant explanation of familiar observations did not involve any experiments, and this probably pleased him. Writing a quarter of a century later about his early work, he says: 'Acute suggestion was then, and indeed always, more in the line of my ambition than experimental illustration.'

In 1801 Young gave two lectures at the Royal Society, in which he applied to optics the principles of interference that he had elucidated in his work on acoustics.[88] He assumed that light waves would, like sound waves, be longitudinal, and though, as we shall see later, that turned out to be wrong, it didn't affect his initial arguments. Again, his important and elegant work did not depend

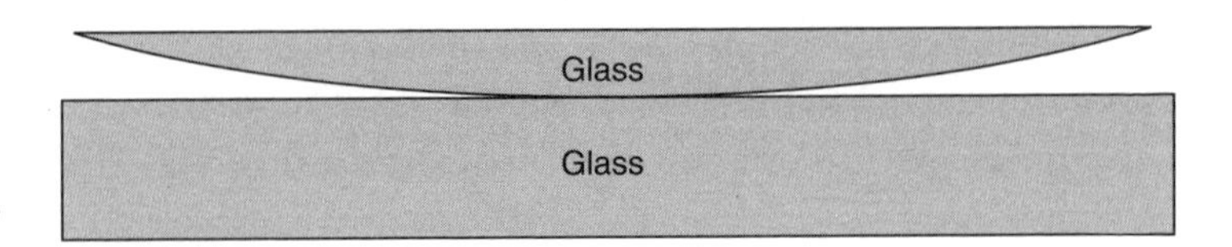

FIG 35 Newton's arrangement for producing a film of air that is thin but of known thickness.

on his own experiments. Instead, he turned to the precise measurements that Newton had made when he was investigating the cause of the beautiful colours seen in thin films—soap bubbles, traces of oil on rainwater puddles, or (less obviously) thin films of air. Newton had placed a thin convex lens of known curvature on a flat glass plate, trapping a film of air between the two, whose thickness at any point could be calculated from its distance from the point of contact between the lens and the plate—see Figure 35. A beam of white light was directed vertically downwards onto the lens. This light was reflected upwards from two surfaces: the upper and lower surfaces of the film of air. Looking down on the lens from above, Newton saw a series of coloured rings centred on the point of contact, the arrangement of colours being that shown in Figure 36. If he looked at the glass plate from below, he saw similar rings but in complementary colours—i.e., the colours which, mixed with the first colours, would give white light. When the lens was illuminated with light of a *single* colour, Newton saw about 30 concentric rings of that colour, and he noticed that *starting in the centre, and passing from one ring to the next, the thickness of the film of air always increased by the same amount.*[89]

Newton assumed, correctly, that the pattern of colours when he used white light was the result of the overlapping of the rings produced by the individual colours present in the white light. The complementary colours seen when the glass plate was viewed from below were also easily explained, since at each point the

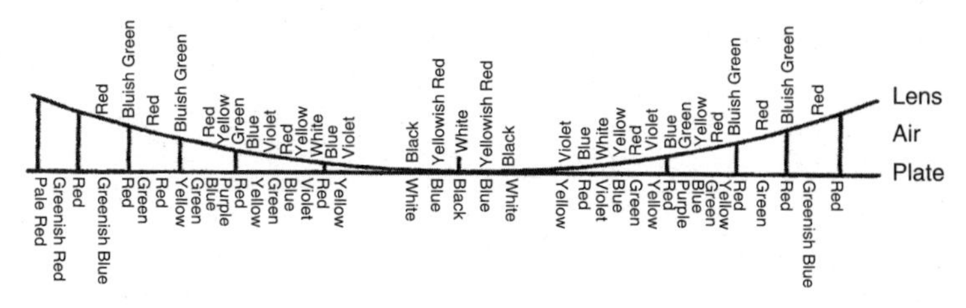

FIG 36 Diagram from Newton's *Opticks* showing the arrangement of colours in Newton's rings.

colour seen from below was simply that of white light from which the light reflected upwards had been subtracted. But he had no satisfactory explanation of the appearance of rings in the first place. As with the problem of simultaneous reflection and refraction when the sun shines on water (see page 117), he invoked the notion that light corpuscles can exist in two states—'fits of easy reflection' and 'fits of easy transmission'—and that motion of the ether may affect these states. But, as Young wrote, 'Even this supposition does not much assist the explanation.'

Young's explanation of 'Newton's rings' was as simple as it was elegant. Looking at any point on the thin film from above, an observer will be receiving beams of light reflected from both the upper and lower surfaces of the film. The total length of the light path—from the light source to the eye of the observer—will be longer for the beam that is reflected from the lower surface than for the beam reflected from the upper surface, the difference being precisely twice the thickness of the film. If that difference in path length is exactly equal to the wavelength of the light, or an exact multiple of that wavelength, peaks of the two beams will reach the eye simultaneously, and so will troughs of the two beams. The beams can be described as 'in phase with one another' and the intensities of the two beams will be added—see Figure 37a. If the difference in path length is exactly equal to *half* the

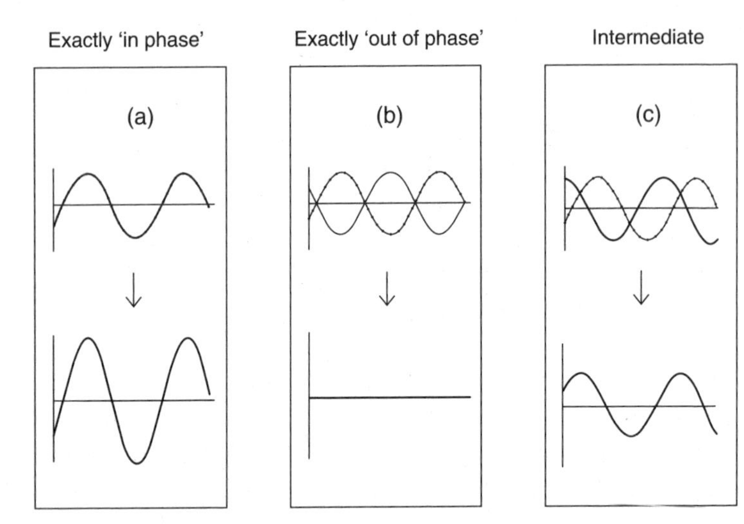

FIG 37 Thomas Young's explanation of the effects of phase difference on the interference between two identical waves.

wavelength of the light, or an odd multiple of that half-wavelength, the peaks of one beam will coincide with the troughs of the other, and vice versa. The two beams can be described as 'exactly out of phase' and no light will be seen—see Figure 37b. (For intermediate differences in path length, the intensity will be less than that seen when the beams are in phase and more than when they are exactly out of phase—see Figure 37c.) As Young put it: if the path difference is equal to or an exact multiple of the wavelength, the *interference* will be *constructive*; if the path difference is equal to or an odd multiple of half the wavelength, the *interference* will be *destructive*.

This not only explains why, with light of a single colour, Newton saw concentric rings; it also explains his observation that 'in passing from one ring to the next, the thickness of the film of air always increased by the same amount.' That increment would be half the wavelength of the light being used. In fact, Young used Newton's data to calculate the wavelengths of light of each of the colours of the spectrum, from red, with the longest wavelength, to violet, with the shortest. These were the first estimates ever made, and, judging by modern data, were surprisingly accurate—though the credit for that must go to Newton rather than Young.

Elegant and persuasive as Young's explanation of Newton's rings was, there was no direct experimental evidence that interference between two beams of light was involved. In 1803, despite his preference for 'acute suggestion' over 'experimental illustration', Young described to the Royal Society an experiment in which he made a hole in a window shutter, covered it with thick paper and then made a pinhole in the paper.[90] A mirror outside the window reflected sunlight horizontally through the pinhole, producing a narrow diverging beam which illuminated a patch on the opposite wall. He found that if he put a 'slip of card about one thirtieth of an inch in breadth' in the path of the beam, fringes of colour appeared on either side of the card's shadow, 'and the shadow itself was divided by similar parallel fringes, of smaller dimensions...but leaving the middle of the shadow always white.' He believed that the fringes in the shadow were the result of interference between light passing the two sides of the card, and to test this he placed a little screen so that it would prevent light passing one side of the card from reaching the wall. As he expected, the fringes in the shadow disappeared. And the disappearance was not simply the result of halving the amount of light falling on the wall, since if he allowed the light to pass on both sides of the card the fringes remained visible even when he reduced the intensity of the light to one tenth or one twentieth.

This experiment was simple and informative, but it was followed by an even more elegant variant that became famous as

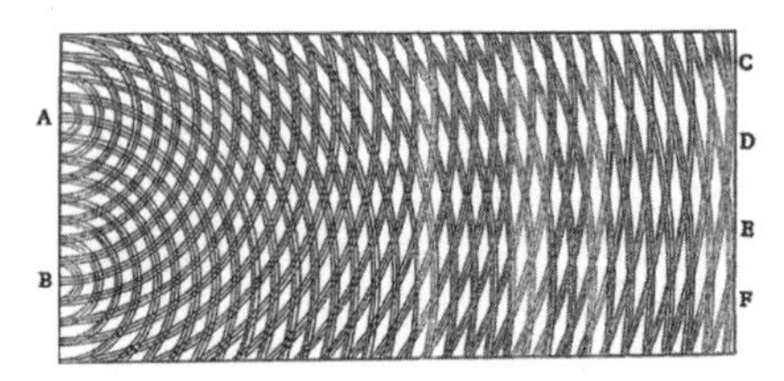

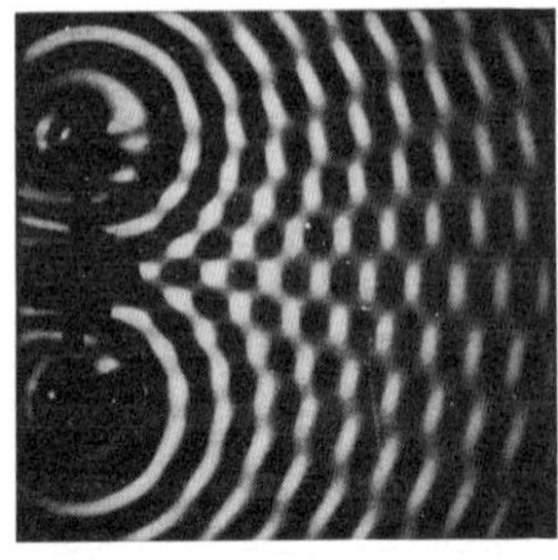

FIG 38 Interference between two equal sets of water waves. The left-hand side is from a drawing by Thomas Young. The right-hand side is a photograph of a modern ripple tank.

'the two-slit experiment'. Young first described it in a lecture at the Royal Institution; he never published the full experimental details, and the first printed version did not appear until the publication of his massive volume *A Course of Lectures on Natural Philosophy and the Mechanical Arts*, in 1807.[91] Instead of using a 'slip of card' to produce two beams of light he arranged that light from a single source and of a single colour fell on a screen in which there were two parallel very narrow slits, close to one another. The diverging, diffracted light coming from the slits fell on a further screen, and where light from the two slits overlapped interference created a series of parallel dark and light bands. Before discussing these bands, Young reminded his audience (or readers) of the pattern of interacting waves that is produced by throwing two similar stones into a pond at the same instant and not too far apart—see Figure 38. He particularly drew attention to the way in which, at some distance from the stones, the interacting waves form a pattern of radiating bands, with the centre of each band showing strong waves and the edges of the band showing little agitation.

Young then turned to the interference pattern produced by light in his 'two-slit' experiment to see how this compared with

FIG 39 The interference pattern observed by Thomas Young using coloured light admitted through two small apertures.

the water pattern. Here, there was a minor though awkward difficulty. In the water experiment, by looking down from above he could see at a glance how the pattern of waves changed as the waves got further and further from their origin where the stones were dropped. In the two-slit experiment, he could only find how the pattern changed with distance by moving the second screen in a series of steps further and further from the slits, and comparing the patterns at successive steps. He found that as the second screen was moved further from the slits, the light and dark bands became wider and wider though always subtending 'very nearly equal angles from the apertures [slits] at all distances.' Figure 39 is a black and white reproduction of a drawing that Young made to show the way the pattern must have changed with distance to account for his observations. The resemblance to the water pattern is striking.

And Young went one step further. Knowing the distance between the screens, the distance between the slits and the positions of the dark and light bands, by repeating the two-slit experiment with light of different colours, and assuming that the difference in path length between successive light bands (or successive dark bands) must always have been one wavelength, he was able to calculate the wavelength of light of each of the colours. The estimates he obtained[92] were in good agreement with his earlier estimates from Newton's rings.

You might think that Young's explanation of Newton's rings, and his experiments looking at the effects of interference between two beams of light, would have quickly persuaded most men of science—the word 'scientist' didn't yet exist—of the truth of the undulatory theory. In fact, it took over twenty years.

There were three reasons.

The first was a natural reluctance to reject Newton's views.

The second was the publication, in the new and very successful *Edinburgh Review*, of three anonymous and monstrously unfair articles dismissing Young's work as 'destitute of every species of merit', asking whether the 'Royal Society has downgraded its publications into bulletins of new and fashionable theories for the ladies who attend the Royal Institution', and describing Young's 'law of interference' as 'one of the most incomprehensible suppositions that we remember to have met with in the history of human hypotheses'.[93] The author was Henry Brougham—later Lord Brougham, lawyer, member of Parliament, Lord Chancellor, supporter of the anti-slavery movement and the Reform Bill, and inventor of the eponymous carriage—but then in his mid-20s and still smarting from criticisms made by Young of a mathematical paper Brougham had published in the Royal Society's *Transactions*. Young ignored the first two articles, but in November 1804, now married and worried about the possible effect of the continued criticism on his developing medical practice, he produced a pamphlet replying vigorously to the third article, naming the writer as Brougham, and ending: 'With this work my pursuit of general science will terminate: henceforward I have resolved to confine my studies and my pen to medical subjects only.' Fortunately he did not stick to his resolution.

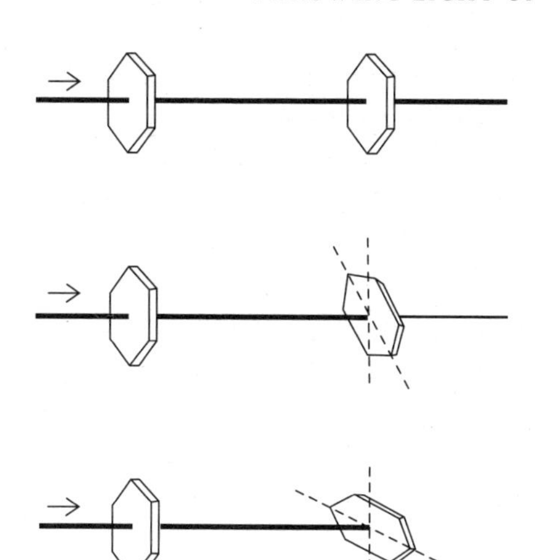

FIG 40 The unexpected effect of orientation on the transmission of light through two crystals of tourmaline.

The third reason was the most serious, and the most interesting, since it led to a deeper understanding of the nature of light. It had been known since the seventeenth century that if you shine a beam of light at one face of a crystal of Iceland spar (a peculiar crystalline form of calcium carbonate) you get *two* refracted rays. If you look at a small object through a flat polished plate of the crystal you see two identical images of the object. Later, equally puzzling results were found with the mineral tourmaline (a crystalline form of a particular borosilicate). Figure 40 shows the effect of shining a beam of light through two thin plates of tourmaline cut with their faces parallel to the axis of the crystal. In the top row, with the axes of the two plates parallel, the beam of light transmitted by the first plate is fully transmitted by the second plate. In the bottom row, with the axes of the two plates at right angles, none of the light transmitted by the first plate is transmitted by the second plate. In the middle row, with an intermediate orientation of the plates, part of the light transmitted by the first plate is transmitted by the second plate. Very similar behaviour can be seen using the lenses of Polaroid sunglasses. If you arrange a pair of these lenses so that one is in front of the other, and they are both oriented as they were in the sunglasses, you will be able to see through them. If you rotate one of them by 90° you will see nothing.

In the first decade of the eighteenth century neither the undulatory theory nor the corpuscular theory seemed to provide any explanation of these strange results. The effect of orientation was particularly puzzling, since if, as supporters of the undulatory theory assumed, light waves were longitudinal waves (like sound waves), they would be radially symmetrical in cross section; so how could rotating the crystal around an axis aligned with the light beam have any effect? (You can see this if you imagine sound waves passing down a pipe whose cross section is any shape you like. What is actually passing down the pipe is a rapid series of transient pulses in the pressure, the molecules of air being moved alternately a very short distance forwards and backwards along the pipe. In this situation, rotating the pipe by 90° around its own axis will have no effect on the passage of the pressure pulses.)

Young was very much aware of these problems but had no answer to them; and after the publication of his *Course of Lectures in Natural Philosophy*, he turned his attention to his medical career. In the course of the next eight years he became a Cambridge MD, a Fellow of the Royal College of Physicians, and a physician at St George's Hospital, and he wrote two books on medicine, both with a somewhat historical slant.[94]

By 1815, though, his enthusiasm for medicine was beginning to wane. In the previous year he had become fascinated by the Rosetta Stone—a fascination which lasted the rest of his life—and in 1817 he returned to the unsolved problem of the behaviour of light in Iceland spar and tourmaline crystals. In a letter to François Arago[95] (whose 'wheel' we met in Chapter 5), he suggested that though light waves were assumed to consist of longitudinal vibrations—that is, vibrations along the line of the ray—they might also have a component of transverse vibration—that is, vibration in directions at right angles to the line of the ray. And if that transverse vibration were restricted to particular directions (say, up and down)—or, as we should now say, 'if the light were vertically polarized'—the radial symmetry would be lost, and the effects of reorienting the crystal transmitting the light could be explained by supposing that the crystal allowed vibrations only in certain directions.

Again, the argument is easier to follow by looking at a mechanical analogy. If you take a thin rope, like a skipping rope, fasten one end to a wall, hold the other end in your hand so that the rope is not quite taut, and rapidly move your hand up and down, a succession of waves will pass along the rope. Each point on the rope will, in turn, move up and down with the same frequency as the movements of your hand. If instead of moving your hand up and down you move it from side to side, a succession of waves will move along the rope, but this time the movements of each point on the rope will be from side to side. By moving your hand appropriately you can also generate waves in which the movements of each point on the rope are in directions intermediate between the vertical and the horizontal—see Figure 33. All these waves involve 'transverse' vibrations, in the sense that the movements of points on the rope are always in directions perpendicular to the path of the waves.

Now imagine that between the wall and your hand the rope passes through a *vertical* slot in a wooden board, the slot being only a little wider than the thickness of the rope. With this arrangement, only vertical waves would be able to pass through the slot. If you rotated the wooden board through 90°, so that the slot was *horizontal*, only horizontal waves would be able to pass through. More generally, transmission beyond the slot is possible only if the orientation of the slot is similar to the orientation of the vibrations. If light waves include transverse vibrations, then, and if the structure of crystals through which the light is passing is such that vibrations are possible in some directions but not others, the kind of behaviour seen with Iceland spar and tourmaline begins to look understandable.

Arago was a friend of another Frenchman, Augustin Fresnel, who in 1815, working in Normandy and knowing nothing of Young's work, had rediscovered Young's principle of interference. Arago told Fresnel of Young's suggestion about a transverse component, and by 1821 Fresnel had shown by careful mathematical analysis that the theory would explain the experimental observations only if all the vibrations in light waves were transverse.[96] Young, who had for so long regarded light waves as analogous to sound waves, took some persuading but was eventually convinced. Young felt that he had planted the tree, and that Fresnel had produced the apple. Fresnel felt that the apple would have been produced even if Young had not planted the tree.[97] But the two got on well, and in due course Fresnel became a Fellow of the Royal Society and Young became a foreign member of the Paris Académie des Sciences.

Arago, incidentally, was also responsible for doing one of the most elegant experiments supporting the wave theory of light, though the idea came from the very distinguished French mathematician Siméon Poisson. Poisson was strongly opposed to the wave theory and he produced a 'thought experiment' that he assumed disproved it. When a beam of light from a distant point source shines on a disc held at right angles to the beam, Huygens' 'construction' says that each point on the circumference of the disc acts as a new point source of light. Some light from these sources will reach the shadow of the disc (on a screen held perpendicular to the source of light), but almost all points in the shadow will be at different distances from different parts of the circumference, so the light from the different point sources will not be in phase. The one exception is the central point in the shadow, which is at an identical distance from every point on the circumference. If the wave theory of light is correct, we should therefore see a bright spot in the centre of the shadow—which Poisson thought was absurd. Arago did the experiment and saw the spot—since then famous as 'Poisson's spot'.[‡]

Now to return to polarized light. A quarter of a century after 'Poisson's spot', Faraday—by then in his mid-50s—showed that when polarized light passing through a glass block was subjected to a strong magnetic field, the plane of polarization of the light was rotated, the angle of rotation being proportional to the strength of the field.[98] This dramatic link between light and electromagnetic theory was followed up, first by Faraday and then by Clerk Maxwell, who by 1864 had produced strong evidence that the transverse waves that make up light are in fact electromagnetic waves;[99] that is to say, the light ray possesses electric and magnetic fields vibrating in directions that are perpendicular to the path of the ray and are also perpendicular to each other—see Figure 41.

With these features, the wave theory provided an elegant and convincing explanation of all aspects of the behaviour of light known at that time; but it turned out that its significance was even wider. For by the beginning of the twentieth century it had

[‡] An argument very similar to that used to explain Poisson's spot will also explain why, when Young noticed parallel fringes of light in the shadow of a very thin slip of card placed in a narrow diverging beam of sunlight coming from a pinhole in the window shutter, he found that the middle of the shadow was always white—see p. 125.

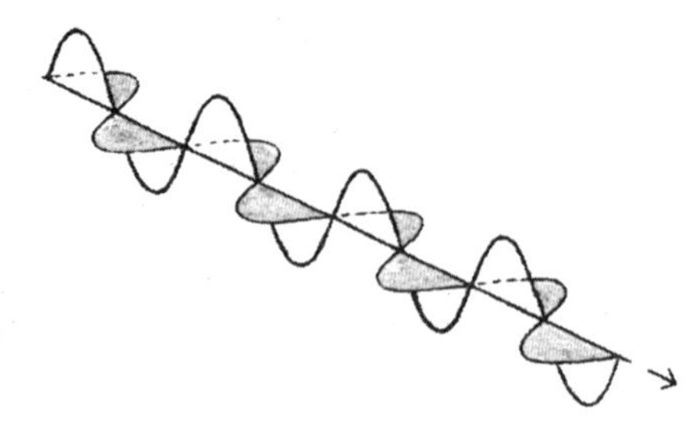

FIG 41 The electromagnetic character of light waves.

become clear that it was only within a narrow band of wavelengths that electromagnetic waves had the properties of light. With longer wavelengths they were radio waves, and with shorter wavelengths they were X-rays or gamma rays—high-energy rays emitted by certain radioactive compounds. But that is another story, and I want to continue with the story of light, for it has a very surprising epilogue.

By the beginning of the twentieth century it looked as though the controversy about the nature of light that had gone on since the seventeenth century was over. The wave theory—the twentieth century preferred the single syllable *wave* to the five syllables in *undulatory*—was generally accepted, and the corpuscular theory seemed to be dead. In the first few years of the century, however, the German physicist Max Planck became interested in the distribution of energy between the different wavelengths of light that were emitted from bodies heated sufficiently to emit light—that is, to become red hot or white hot. He found an empirical formula that fitted the observed distribution, but when he tried to explain that distribution on classical wave theory, he found that he couldn't. To explain it, he needed to assume that when hot bodies emit energy as radiation, the energy is supplied in discrete packets (later known as quanta), the size of the packet determining the wavelength of the light emitted—with the shortest wavelengths being formed by the packets with most energy. He did *not* suggest that the radiation itself was 'quantized', believing that the waves emitted were simply the classical electromagnetic waves of Maxwell.[100]

But in 1905 Einstein went further, claiming that light was *transmitted* in quanta whose size determined the wavelength. This looked embarrassingly like a reversion to the corpuscular theory, and initially found little support.

An important part of his argument was that the hypothesis that light was transmitted in quanta could explain the puzzling features of the production of an electric current by the action of light on a photoelectric cell. Today these cells are used in all sorts of devices, but the basic idea was known at the beginning of the twentieth century. There are certain materials, such as sodium, potassium, and caesium, which tend to emit electrons when light falls on them. A photoelectric cell consists essentially of an airless glass tube containing two electrodes, one—the emitting electrode—with its surface covered with, say, a thin layer of potassium. When light falls on the emitting electrode some electrons gain sufficient energy from the light to escape from the metal and form a cloud of negative charge around it. Some of the electrons with the greatest kinetic energy will move across the vacuum to the other electrode—the collecting electrode—and a galvanometer connected between the two electrodes will register a current. It is possible to measure the kinetic energy of the most rapidly moving electrons by seeing how big a negative voltage has to be applied to the collecting electrode to stop this current. An unexpected finding, by a German physicist, Philipp von Lenard, was that the number and the maximum energy of the escaping electrons depended in peculiar and surprising ways on the wavelength and intensity of the incident light. Firstly, although it was known that different emitters responded differently to light of different wavelengths, it appeared that for any given emitter there was a threshold wavelength beyond which no electrons were emitted *however intense the light*. (That was odd, since you would expect that whatever energy was needed to dislodge an electron and give it enough kinetic energy to reach the other electrode, would be available if you increased the intensity of the light enough.) Secondly, for a given emitter and light of a given wavelength (within the effective range), the *number* of electrons emitted per second was, as you might expect, proportional to the intensity; but the *maximum energy* of the emitted electrons was totally unaffected by the intensity. (This too was odd: why should the energy of the released electrons not go on increasing as you increase the intensity of the light?)

Einstein realized that these curious findings could be explained by adopting the hypothesis that light comes in packets of energy (quanta) whose size is constant for light of any particular wavelength, and varies inversely with the wavelength when light of different wavelengths is compared. He suggested that the emission of each electron from the emitting electrode was the result of the transfer to it of the energy in one quantum of light, and that that transfer was an 'all-or-none' event. Either the entire energy in the quantum was transferred, and the quantum disappeared, or none of it was transferred and both the quantum and the electron remained as they were. To dislodge an electron from the potassium (or whatever the emitting material was) needed a certain amount of energy, so if the energy in the quantum was insufficient no electrons would be released however intense the light. That explained the threshold.

Dislodging an electron from the metal would be easiest if the electron was at the surface, but it would still need a certain amount of energy—what physicists called the *work function*. The maximum kinetic energy that a released electron could have would therefore be the amount in the quantum less the *work function*. For light of any given wavelength, the amount of energy in a quantum is fixed. Increasing the intensity of the light will, of course, increase the number of quanta available, but it won't alter the amount of energy in each quantum, nor will it alter the *work function*. That is why increasing the intensity of the light increases the number of electrons emitted, but it doesn't alter the maximum kinetic energy that those electrons can have.

Einstein's interpretation was brilliant—he would eventually get a Nobel Prize for it, not for his work on relativity—but it was also very disturbing. Even Planck was disturbed. The corpuscle/wave argument that seemed to have been settled for nearly a century was suddenly reanimated. And the position soon got much worse.

In 1908 Geoffrey Taylor, later to become extremely distinguished in the field of fluid mechanics, had just graduated as a BA in Cambridge and was ready to start research. J.J. Thomson, the discoverer of the electron, and at that time Cavendish Professor of Experimental Physics, suggested a number of possible lines, and the one chosen by Taylor was this: if, as has been suggested, light consists of quanta of energy localized in space, interference must depend on more than one quantum falling on the same spot at the same time, or almost the same time. If, then, the intensity of light forming the interference pattern is reduced, the average time interval between successive quanta falling on the same spot will be increased; so by making the light weak enough, it should be possible to make the normal interference pattern disappear.[101]

To test this prediction, Taylor modified the technique that Young had used in 1803 in those experiments in which interference patterns had been produced by inserting a thin slip of card in a beam of sunlight diverging from a pinhole. Here is Taylor's description of his own experiment:

> I...set up in the old children's playroom at my parents' home a vertical needle on to which I shone the light of a gas jet placed behind a vertical slit made by a razor blade in a piece of metal foil stuck to a piece of glass. I set a photographic plate up some six feet away and obtained a good interference pattern. Then I reduced the light by inserting successive sheets of dark glass between the gas jet and the slit and increased the exposure time.[102]

He reckoned that with the darkest glass the intensity of the light falling on the screen was similar to that that would be caused by a standard candle burning at a distance of about one mile. Yet using this glass, after three months exposure—i.e., a period long enough for sufficient light to have reached the photographic plate to darken it—he obtained interference patterns as clear as those obtained in a few minutes with a stronger light.[103]

This persistence of interference patterns however weak the light (subsequently confirmed by others using Young's later two-slit method) was described by Richard Feynman as 'a phenomenon which was impossible, *absolutely impossible*, to explain in a classical way'.[104] With extremely weak light, very few of the photons will pass through the slit (or two slits) simultaneously. We know from Young's experiments that interference requires the interaction of light from both slits (or from both sides of the slip of card in the earlier experiments) but how can a single photon go through two separate slits at the same time? It was the impossibility of explaining the results of these simple and elegant experiments 'in a classical way' that led to the development of the whole field of quantum mechanics.

As far as light is concerned, we have to accept that it can behave both like a succession of waves, and like a succession of particles. We also have to accept that when we are considering events on a scale enormously smaller than anything we meet in ordinary everyday life, things may behave in ways that are not intuitive, and are not always in line with 'laws' derived from studies at larger scales. The laws of quantum mechanics are extremely successful in describing and predicting events, but they are strange; and why they are what they are is not understood. It is this that accounts for Feynman's remark, in 1965, that he thinks he can 'safely say that nobody understands quantum mechanics'.[105] My physicist colleagues tell me that that remark remains true today.

7

HOW DO NERVES WORK?

From the time of Galen (born AD 129) until well into the seventeenth century, the standard answer to the question: 'How do nerves work?' was that nerves produce their effects through a flow of 'animal spirits'. As animal spirits were thought to be weightless, intangible, and invisible—a bit like the later and equally mythical 'caloric'[106]—to say that nerves produce their effects by the flow of such spirits was to say little more than that nerves produce their effects by the flow along them of something otherwise undetectable. This was, of course, true but not very helpful. By the middle of the twentieth century the problem had largely been solved. The story of its solution, over a little more than three centuries, is complicated, but it involves a number of elegant experiments and hypotheses, and it is some of these that I want to discuss in this chapter. I shall not talk about the way the brain works, or the way sense organs work, but only about the more limited problem: how do nerves transmit information? To appreciate any experiment or any hypothesis, though, you need to know the state of play at the time the experiment was done or the hypothesis was proposed; so from time to time I shall try to provide some sort of background.

In the seventeenth century, Descartes suggested that animal spirits might be a real fluid behaving according to the same physical laws as other fluids, and that it was the swelling of muscles by animal spirits conveyed to them through hollow nerves that caused them to shorten.[107] He even believed that the nerves had valves. Giovanni Borelli, a contemporary of Descartes and Professor of Mathematics at Pisa, thought it unlikely that there was a sufficient flow of fluid to inflate the muscle directly, but he suggested that the release of a few drops of the 'spirituous juice' from the nerve could initiate 'a sudden fermentation and ebullition' in the muscle, so 'filling up the porosities' and inflating it.[108] Both of these interesting theories were demolished at a stroke by a simple and elegant experiment by Jan Swammerdam, an Amsterdam physician a generation younger than Descartes and Borelli, who did very little medicine and spent most of his time studying biology.

The crucial experiment is illustrated in Figure 42.[109] The calf muscle of a frog, with its attached nerve, was suspended in a glass vessel closed at the bottom by a cork. A thin wire passing upwards through the cork was looped around the nerve in such a way that pulling on the wire pinched the nerve, exciting it, and causing the muscle to twitch. A fine vertical glass tube, with a drop of water trapped in the middle, emerged from the top of the vessel. If, for whatever reason, the muscle swelled when it contracted,* the pressure of air in the vessel would increase and the drop of water would move upwards. It didn't.

Oddly, although this experiment was performed by Swammerdam in the 1660s, and was demonstrated to his contemporaries, it was not published until more than fifty years after his death, when Boerhaave—the man whose chapter on fire had

* Since *to contract* is sometimes used to mean *to decrease in volume*, or *to shrink*, it may seem perverse to suggest that a muscle increases in volume when it contracts. But the English word comes from the Latin *con trahere*, meaning literally *to draw together*, and it is of course by drawing their two ends together that muscles produce movement.

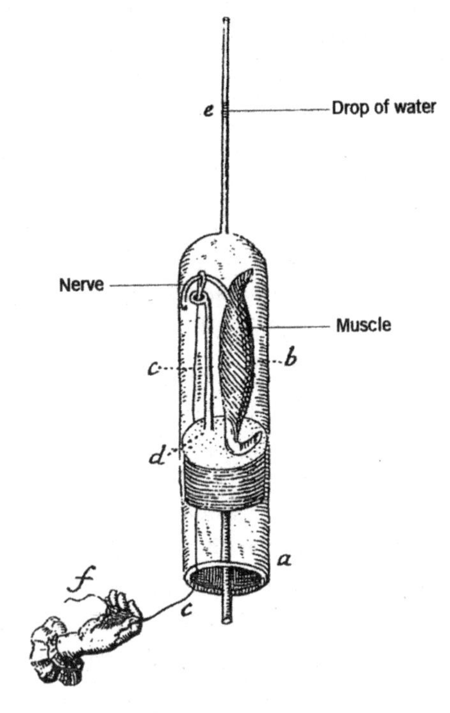

FIG 42 Swammerdam's experiment showing that a muscle does not change its volume when it is made to contract.

inspired Rumford—published *Biblia Naturae*, a collection of Swammerdam's works in Dutch with Latin translations.

By the end of the eighteenth century, the notion that nerves work by transmitting 'animal spirits' was widely recognized as worthless; but there was no satisfactory hypothesis to replace it. It was known from the work of Galvani and others that frog nerves could be excited by quite weak electric shocks; and various fish—electric catfish in the Nile, torpedos (or stingrays) in the Mediterranean, and electric eels in South American rivers—were known to cause numbing shocks that were thought to be electrical. But it was not at all clear that electricity was involved in the normal process of transmitting messages along nerves.

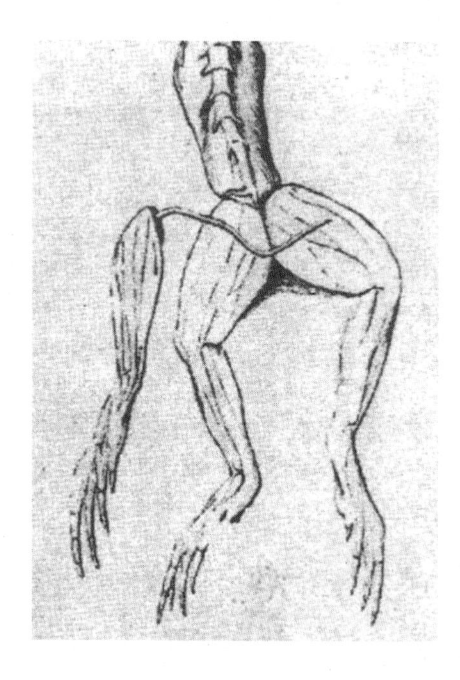

FIG 43 Matteucci's 'induced twitch' experiment.

In the 1830s and 40s, Carlo Matteucci, who had taken a physics degree at Bologna, and who in 1840 was appointed Professor of Physics at Pisa, did two simple but very informative experiments.

In the first, which became known as the 'induced twitch' experiment, he draped the sciatic nerve of the detached leg of a recently killed frog across the thigh muscle of another recently killed frog (see Figure 43). When the thigh muscle of the second frog was made to contract by either an electrical or a mechanical stimulus, the leg of the first frog also twitched.[110] The most plausible explanation (which turned out to be correct, though later in his life Matteucci doubted it) was that electrical changes on the surface of the contracting muscle excited the nerve draped across it.

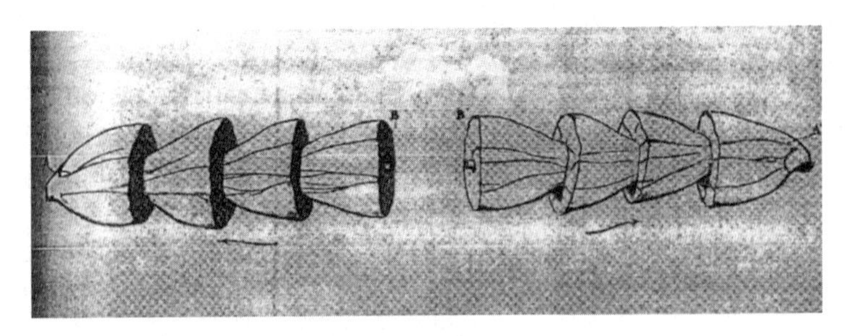

FIG 44 Matteucci's battery of half-thighs.

The second experiment took advantage of a recent improvement in the sensitivity of galvanometers, to show that if you took the thigh muscle of a recently killed frog, cut it across transversely, and put one wire from a galvanometer on the cut surface and the other on the intact outer surface, you would detect a small steady current, later referred to as the 'injury current'.[111]

Matteucci's preferred explanation was that there is normally a difference in electrical potential (voltage) between the inside and the outside of the resting muscle, but there was a possible alternative explanation. During the last part of the eighteenth century, Galvani had performed several experiments that seemed to demonstrate the existence of voltage differences in living tissues, but which Volta later showed were the result of reactions that may occur when two dissimilar metals are in contact. Could it be that the 'injury current' observed by Matteucci was also an artefact, driven by a voltage caused by a reaction between the wires and the muscle tissues? To test this possibility, Matteucci built 'batteries' of half-thighs, as shown in Figure 44, and compared the size of the galvanometer deflection with the number of half-thighs in the battery.[112] It turned out that the deflection increased roughly in proportion to the number of half-thighs. Since the number of metal–tissue

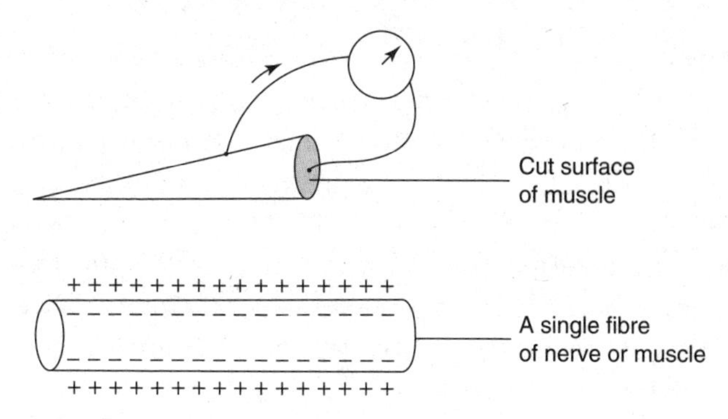

FIG 45 Du Bois-Reymond's interpretation of the 'injury current'.

contacts remained the same, this was game, set, and match to Matteucci's view.

The observation that an 'injury current' could be detected if a galvanometer were connected between the cut surface and the intact outer surface of a muscle was confirmed by Emil Du Bois-Reymond, who went on to show that similar injury currents could be shown in nerves or in small bundles of muscle fibres. He assumed that this was because the cut surface allowed access to the interior of the individual fibres of the nerve or muscle, and from the direction of the current he concluded that the interior of the normal resting nerve fibre or muscle fibre is at a lower electrical potential (lower voltage) than the fluid outside it—see Figure 45. This was in fact correct, and the *difference in potential* across the membrane of the inactive nerve or muscle fibre became known as *the resting potential.*

The crucial question, though, was: 'What, if anything, has this difference in electrical potential to do with the transmission of signals along nerves, or the spread of excitation in a muscle?' One way of approaching this problem was to ask: 'What happens to the injury current when the nerve is busy transmitting signals or when the muscle is being stimulated to contract strongly?' Working first with frog muscle, Du Bois-Reymond therefore looked at the effect on the injury current of making the muscle contract violently with a rapid series of small electric shocks to the nerve controlling the muscle. He found that the injury current decreased or even disappeared—a change he referred to as 'the negative variation'. Turning then to nerve, he found a similar effect. It seemed that the spread of excitation through the muscle, or the transmission of signals along the nerve, was accompanied by a transient decrease in the potential difference across the membrane. This transient change later became known as *the action potential*. Du Bois-Reymond himself had no doubt about the significance of his findings:

> If I do not greatly deceive myself, I have succeeded in realizing in full actuality (albeit under a slightly different aspect) the hundred years' dream of physicists and physiologists, to wit, the identity of the nervous principle with electricity.[113]

At first sight, this claim might have seemed a shade more lyrical than logical, since it was well known that a crushed nerve or a ligatured nerve could continue to conduct electric currents though it could not transmit signals. But Du Bois-Reymond suggested that the resting nerve membrane contained an orderly array of roughly spherical molecules, positively charged in their 'equatorial zones' and negatively charged at their 'poles', and that an electrical stimulus to the nerve caused a reorientation of these molecules leading to the 'negative variation'.[114] Such an arrangement could well be upset by crushing the nerve or putting a ligature round it. This of course was just speculation but, considering his work as a whole, a big step had been taken. And that step pointed to a possible next step.

Possible, but not easy to take. If the transient drop in the potential difference across the membrane of the nerve fibre was indeed part of the mechanism by which signals pass along the nerve, you would expect the area of membrane showing the drop to pass along the nerve at the same speed as the signal. Did it? Answering that question depended on measuring the speed of each. And at a time when most people believed (wrongly) that nerve signals passed along nerves immeasurably fast—perhaps at a speed comparable with that of light—and galvanometers were so sluggish that it could take more than 10 seconds for the magnetic needle to reach a steady deflection for any given current, the task looked formidable. Fortunately Herman von Helmholtz, 28 years old, newly married and working in Königsberg, recognized that, although common sense suggested that the speed at which messages pass along nerves must be pretty fast, there was no sound basis for supposing that it was immeasurably fast. He may also have been influenced by Du Bois-Reymond's model suggesting that the spread of excitation along a nerve fibre involved reorientation of molecules in the nerve membrane. Anyway, he decided to ignore the prevailing view, and to attempt to measure the rate at which the signal initiated by a single momentary electric shock to the sciatic nerve of a recently killed frog** passed along the nerve to the calf muscle, where it caused a brief contraction (a twitch).

** The popularity of the frog's sciatic nerve for this sort of experiment is the result not only of the availability of frogs and the ease of their dissection, but also of the ability of frogs' nerves to survive and continue working for several hours at room temperature provided they are kept moist with suitable dilute salt solutions. And it was of course Galvani's observation of the contraction of frogs' legs on contact with dissimilar metals that led to the discovery of current electricity.

To measure the rate at which the signal passed along the nerve it was not sufficient simply to measure the time interval between the initiating shock and the start of the twitch. There might be a delay between the shock and the start of the signal, or between the arrival of the signal at the muscle and the start of the twitch. But if you first measured the overall time interval when the nerve was stimulated fairly near the muscle, and you then compared that with the time interval when the nerve was stimulated much further from the muscle, you would expect the initial and final delays to be the same, so any difference in the overall time would be the result of the extra time it took the signal to travel the extra distance. (That argument assumes that the signal does not change as it goes along the nerve, but since the way the muscle responded was the same irrespective of where along the nerve the initial shock was given, a change seemed unlikely.)

Helmholtz's main problem was to measure a very short time interval with a very sluggish galvanometer, and he did this by an ingenious adaptation of an elegant technique that had been invented by Claude Pouillet, a Professor of Physics at the Ecole Polytechnique in Paris. Pouillet had been interested in measuring the interval between the moment when the hammer of a gun strikes the firing cap and the moment the ball emerges from the gun's barrel.[115] The basis of his technique was to make a virtue of necessity, and take advantage of the galvanometer's sluggishness. He had found that for *steady* electric currents *whose duration was very small compared with the galvanometer's response time*, the maximum deflection of the needle (which of course occurred long after the current had stopped) was related to the duration of the current.[†] He therefore arranged that contact between the hammer and

[†] This makes sense, since with a steady current the total amount of electric charge carried round the circuit would be determined by the duration.

the firing cap completed a circuit that included a battery, a galvanometer, and a length of wire that passed just in front of the mouth of the barrel. When the gun was fired, current would flow through the galvanometer from the moment the hammer hit the firing cap to the moment the ball emerged from the barrel and destroyed the wire. The longer the interval, the greater the deflection of the galvanometer. To measure brief intervals of time, then, he first needed to calibrate the galvanometer, which he did by measuring the maximum deflection after brief shocks of known duration. (These shocks were obtained by having a disc spinning at a known speed, which delivered current only when a sliding contact crossed a radial metal strip fixed to the disc.)

Helmholtz, like Pouillet, used the galvanometer purely as a device for measuring very short intervals of time; it was not directly involved in detecting electrical changes in the nerve or muscle. The galvanometer was included in a circuit that also included a battery and two switches. Helmholtz arranged that the first switch would be closed at precisely the moment at which a very brief but powerful shock (from an induction coil in a quite separate circuit) was given to the sciatic nerve of a recently killed frog. He also arranged that the second switch, which was closed before the experiment began, would be opened as soon as the muscle began to shorten. The result of this arrangement was that a current (driven by the battery, and unrelated to any current in the nerve or muscle) would flow through the galvanometer from the moment the sciatic nerve was stimulated to the moment the muscle began to contract. Since between those two moments the voltage driving current through the galvanometer was constant, the maximum deflection of the needle was determined by the duration of the current. In fact, Helmholtz found that for deflections up to about 20°, the deflection was proportional to the duration. By fixing a tiny mirror to the needle and shining a light at it that was reflected onto a scale, he ensured that measurements of the deflection could be extremely accurate. It was, incidentally, his wife, now pregnant with their first child, who noted and recorded many of the deflections. As Helmholtz explained to Du Bois-Reymond, he himself became 'completely confused when I am supposed to pay attention to so many things at the same time.'[116]

The first successful experiments were done just before and after Christmas in 1849. By early 1850, Helmholtz had clear evidence that moving the point of stimulation from a point on the nerve near the muscle to a point 50 to 60 mm further away increased the interval between stimulation of the nerve and contraction of the muscle by about 1.4–2.0 milliseconds. Later in the same year he published a full account of his findings.[117] For each experiment, he related the distance between the two points of stimulation to the resulting change in the interval between stimulation of the nerve and contraction of the muscle. It was clear that the signal must move along the frog sciatic nerve with a speed of about 30 metres per second; different experiments gave a range of 24.6–38.4 metres per second. Far from being comparable with the speed of light, the speed of conduction of signals along the sciatic nerve of a frog was only about one tenth of the speed of sound.

How did this compare with the speed at which the *action potential*—the transient decrease in the electrical potential difference across the membrane of an excited nerve—moved along the sciatic nerve of a frog? Du Bois-Reymond attempted to measure this speed but failed. He passed the problem to his student Julius Bernstein, but it was not until 1868—eighteen years after

Helmholtz had measured the speed of the signal—that Bernstein, now working in Helmholtz's laboratory in Heidelberg, succeeded. Measuring the speed at which the action potential moved was even trickier than measuring the speed of the signal that caused the muscle to contract, since the only way of knowing when the action potential arrived at a particular point on the nerve was to monitor the membrane potential at that point; and the only way of measuring the membrane potential was with an extremely sluggish galvanometer.

Bernstein's solution was elegant: breathtakingly simple in theory, though needing considerable ingenuity to put into practice.[118] If you want to know what the potential difference across the nerve membrane is at a particular point on the nerve and during a very brief moment, and the only instrument you have with which to measure potential difference is a very sluggish galvanometer, you must arrange things so that the wire from that point on the nerve is connected to the galvanometer only during that brief moment. However sluggish the galvanometer is, so long as it is sensitive enough to respond at all, its response will depend on the electrical potential during that moment. If, further, you want to know how the potential at a particular point on the nerve changes with time following stimulation of the nerve at some distant point—in other words if you want to detect the arrival of the *action potential*—you will need to look at a succession of brief moments each chosen to start at a different interval after the stimulation. Having established what the pattern of change is, you can then repeat the procedure, but stimulating the nerve at a point nearer to the point at which you are recording. You will find that the pattern is the same but the delay is slightly reduced, and by comparing the reduction in the delay with the reduction in the distance from the point of stimulation you will be able to calculate the speed at which the action potential moves along the nerve.

How practicable was this programme? If the action potential moves along the nerve at the same speed as the signal that eventually makes the muscle contract—which is the hypothesis Bernstein was trying to test—its speed must be about 30 metres per second, or 30 millimetres per millisecond. It would therefore take only about two milliseconds to traverse the whole length of the sciatic nerve. To follow the movement, then, the 'brief moments' we talked about in the last paragraph would have to be very brief indeed. Bernstein aimed to make them about a third of a millisecond.

So how did he do it?

Three things were required: a *stimulating system* for delivering a brief electric shock to stimulate the nerve; a *recording system* for measuring changes in the potential difference between any chosen point on the surface of the nerve and the cut end; and a *timing system* for controlling the timing of the stimulating shock and the start and duration of the period during which the galvanometer was connected.

The first requirement—a stimulating system—presented no problems. All that was needed was an induction coil, a battery, and a switch.

The second requirement—a recording system—also seemed simple initially; you would think that a galvanometer and two pieces of wire would do. But there was a snag. Because galvanometers had an extremely low electrical resistance, connecting a galvanometer between the surface of a nerve fibre and its cut end was in effect short-circuiting the nerve membrane. This greatly reduced the membrane potential in the neighbourhood of the contact point between the wire and the surface of the nerve fibre, and this reduction excited the nerve. Bernstein therefore introduced, into the galvanometer circuit, a battery and an arrangement of resistances that provided a voltage that exactly balanced the voltage generated by the resting nerve membrane. With this arrangement, when the nerve was at rest, no current flowed through the galvanometer. After the nerve was stimulated, the arrival of the action potential disturbed the balance and the galvanometer responded.

The major problem for Bernstein was the control of timing. What he needed, in effect, was a conductor controlling an orchestra of three players: one operating the switch in the circuit producing the stimulating shock; another connecting the galvanometer to the recording circuit at some variable but known time after the stimulation, and the third disconnecting the galvanometer at the end of a specified monitoring period. He achieved this by employing not a conductor but a 'musical box'—a horizontal brass wheel that was made to rotate at an accurately controlled speed by an electric motor. At different points near the edge of the wheel, pins projected downwards and operated switches placed below the wheel. By adjusting the relative positions of the switches, he could control accurately and easily the delay between the stimulating shock and the start of the period during which the galvanometer was monitoring events, and also the duration of that period.

To get adequate time resolution, he had decided to reduce the monitoring period to about a third of a millisecond, but making the period so short meant that the flow of current during a single period could have only a very small effect on the galvanometer needle. He overcame this difficulty by allowing the wheel to rotate continuously so that the nerve was stimulated repeatedly and the resulting action potentials were each monitored for a third of a millisecond *at the same point on the nerve and at precisely the same interval after a stimulus*. Because the galvanometer responded so slowly, the effects of the individual samples were cumulative and (for a fixed speed of rotation) the final position of the galvanometer needle reflected the average magnitude of the membrane potential during the successive brief periods monitored by the galvanometer.

By measuring the potential at a fixed point on the surface of the nerve and varying the delay between stimulation and the start of the monitoring period, Bernstein was able to calculate the way in which the electrical potential across the nerve membrane varies with time as the action potential passes through that point.

Using this ingenious and elegant technique, he compared the delays between the stimulus and the beginning of the action potential when the nerve was stimulated at two different distances from the recording point. From the difference in the delays, he was able to calculate the speed at which the action potential travelled down the nerve. It was 28.7 metres per second, in good agreement with Helmholtz's estimate of about 30 metres per second for the speed of the signal that caused the muscle to contract.

The fact that the action potential travelled along the nerve at a speed close to that of the signal being transmitted, strongly suggested that the two were either identical or closely connected. But why does the action potential travel along the nerve at all? A few years after the publication of Bernstein's experiments, Ludimar Hermann, another former student of Du Bois-Reymond, suggested an ingenious and highly plausible explanation that became know as *the local-circuit hypothesis.*[119] Since the electrical potential difference across the nerve membrane in the area experiencing the action potential is much less than the potential difference across the membrane of the resting nerve, it follows

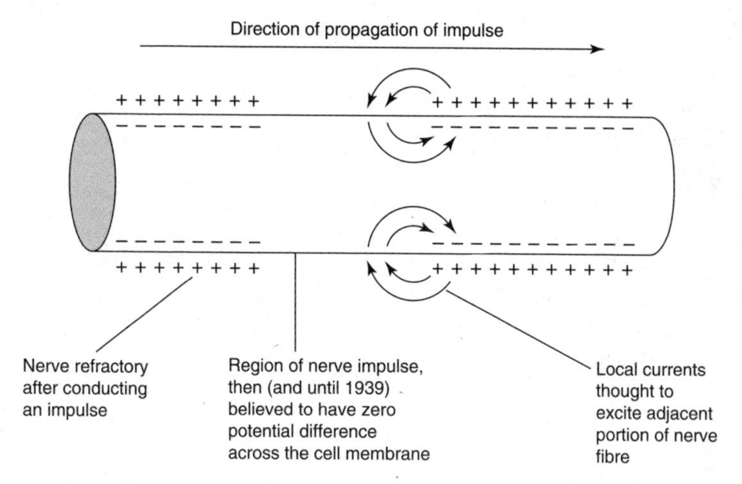

FIG 46 Ludimar Hermann's 'local-circuit hypothesis'.

that at the junction of the active and resting regions of the nerve you will expect to get local electric currents—see the curved arrows in Figure 46. These currents will tend to reduce the electrical potential difference across the membrane of the resting nerve close to the active region, and Hermann suggested *that this reduction in potential difference excites that portion of the nerve*. If that is right, each bit of nerve will become active in turn, excited by local currents from the preceding bit, and producing currents capable of exciting the subsequent bit. Sixty-seven years later, a few months before the Second World War, this ingenious hypothesis would be proved by one of the most elegant experiments in physiology, but we shall come back to that later.

First, we must turn to even more basic questions. Why is there a difference in electrical potential between the inside and outside

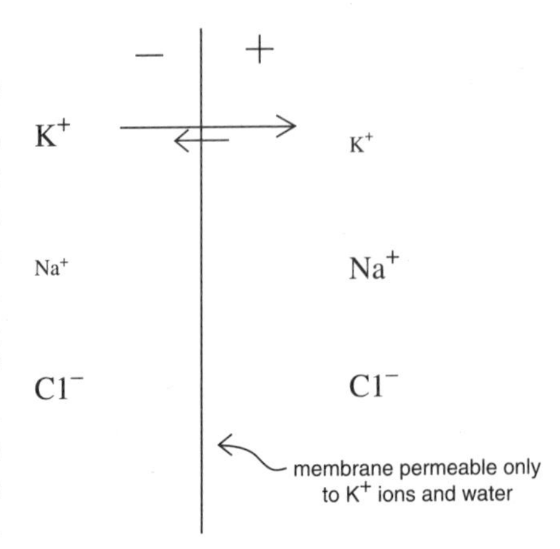

FIG 47 The principle of Wilhelm Ostwald's experiments showing that when a membrane that is selectively permeable to one kind of ion separates solutions containing different concentrations of that ion, an electrical potential difference is generated across the membrane.

of a nerve fibre, and why does that potential change during the transmission of a signal? Or putting the questions another way: why is there a resting potential and what causes the transient changes we call the action potential?

In 1890 Wilhelm Ostwald, a German physical chemist working at Leipzig, studied the properties of artificial membranes that were selectively permeable, allowing some ions to pass through but excluding others. He found that when the fluids bathing the membrane contained only one kind of ion that could pass through the membrane, and ions of this kind were in different concentrations on the two sides of the membrane, there was a difference in electrical potential between the solutions on the two sides.

You can see why this is to be expected if you consider the situation summarized in Figure 47. Here a membrane *permeable only to potassium ions and water* separates two solutions. That on the left contains mainly potassium chloride and a little sodium chloride; that on the right contains mainly sodium chloride and a little potassium chloride. If the total salt concentration is the same on both sides, there will be no movement of water, and initially each solution will be uncharged because both sodium chloride and potassium chloride contain equal numbers of positive and negative ions. But now consider the movements of the potassium ions across the membrane.

Because they are in much higher concentration on the left of the membrane than on the right, and they are moving randomly, far more will cross from left to right than from right to left. Because they are positively charged, this net movement will build up an electrical potential difference across the membrane, positive on the right. As this potential builds up it will become increasingly difficult for a positively charged ion to move to the right, and increasingly easy for it to move to the left. Eventually, the difference in electrical potential will just balance the effect of the difference in concentration, and there will be no further net movement of the penetrating ions. The two solutions will be in equilibrium. In fact, in the situation we have been considering, equilibrium is attained almost instantaneously, and the net amount of potassium moved across the membrane is so small that you would be unable to detect a change in the concentrations on the two sides. In the late 1880s, Ostwald's younger colleague, Walter Nernst, studied situations of this kind and proved that at equilibrium the electrical potential must be proportional to the logarithm of the ratio of the concentrations of the potassium ions on the two sides of the membrane.

How is all this relevant to nerves? In 1902, thirty-four years after he had measured the speed of the action potential in the frog sciatic nerve, Julius Bernstein published another equally famous paper, in which he brought together Nernst's concept of equilibrium potentials across selectively permeable membranes, and Hermann's 'local-circuit hypothesis'.[120] It was already known that the fluid inside animal cells (including nerve fibres and muscle fibres) was rich in potassium ions whereas the extracellular fluid was rich in sodium ions. Bernstein put forward a hypothesis that made three assumptions. *Firstly*, that the membrane of a resting nerve fibre is selectively permeable to potassium ions; *secondly*, that the potential difference across the membrane of a resting nerve fibre is roughly a potassium equilibrium potential (in line with Ostwald and Nernst's ideas); and *thirdly*, that the action potential is the result of a *transient* loss by the membrane of its selective permeability to potassium ions, owing to a general increase in its permeability to other ions, including sodium ions. This hypothesis became known as 'the membrane theory of nerve conduction', and—as we shall see—it turned out to be right in principle but wrong in detail.

First, though, we must turn for a moment from Germany to England, and from considering the machinery for transmitting messages to considering the sorts of message that are transmitted. In 1899 Francis Gotch and G.J. Burch, in Oxford, showed that if the isolated sciatic nerve of a frog was stimulated with two successive shocks, the second shock, however strong, was ineffective if it occurred within a few milliseconds of the first.[121] The exact length of the critical period depended on the temperature, but at 20°C it was about 2 milliseconds. It seems that for a brief period following activity the nerve becomes totally inexcitable or 'refractory'—from the Latin word for stubborn. This refractory period has important consequences. It implies that if it is the action potentials that form the message, any message transmitted along the nerve must consist of a number of discrete 'impulses', which cannot recur at more than a certain frequency. As E.D. Adrian[122] put it, writing nearly thirty years later, 'The nervous message may be likened to a stream of bullets from a machine gun; it cannot be likened to a continuous stream of water from a hose.'[123]

An even more striking limitation of the character of the messages carried by nerves was discovered in 1909 by Keith Lucas in Cambridge.[124] It had been known for some years that by varying the strength of the electric shocks given to the sciatic nerve of a frog it was possible to vary the strength of the twitch produced in the calf muscle—a progressive increase or decrease in the strength of the shock leading to a corresponding increase or decrease in the size of the twitch. But because the sciatic nerve contains a few thousand individual fibres, there was no way of knowing whether this was because the stronger shocks caused a greater response in each individual nerve fibre or simply stimulated a greater number of fibres—or possibly both. Keith Lucas found an elegant solution to this apparently intractable problem by doing a similar experiment but using, instead of the calf muscle, a very small back muscle that was controlled by a nerve that contained only about nine individual nerve fibres. He found that as he increased the strength of the shock to the nerve, the size of the muscle twitch increased, but instead of increasing smoothly it increased in a stepwise fashion—see Figure 48.

What is more, there were never more steps than there were individual fibres in the nerve. The obvious explanation was that each fibre was responding in an 'all-or-none' fashion—either there was a full response or no response; an action potential was transmitted or no effective signal was transmitted. The message along a single nerve fibre was like a stream of bullets not only in its discontinuity but also in that the size of each unit did not vary with the size of the stimulus that initiated it—the pressure of the finger on the trigger—provided that the stimulus was not too small to produce any response.

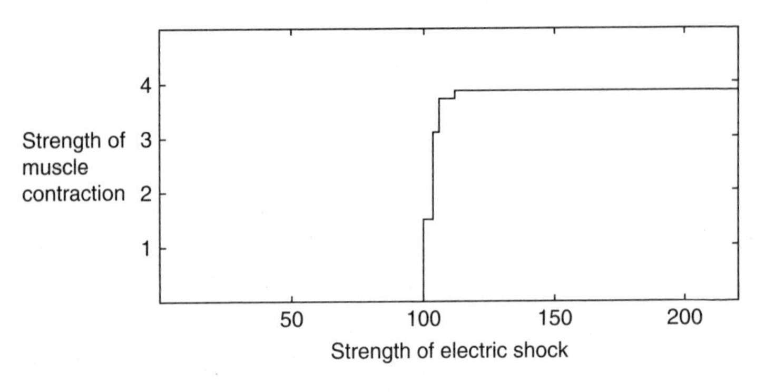

FIG 48 Keith Lucas' experiment showing the 'all-or-none' law.

Although Gotch and Lucas used only electric shocks as stimuli and worked only with motor nerves, later work showed that both the 'all-or-none' character and the refractory period—the brief period of inexcitability following an action potential—were general features of nerve conduction. Nerve fibres, it was clear, are not like traditional telephone wires, which carry a continuously varying current reflecting the continuously varying pitch and loudness of the speaker's voice. They are not even like traditional telegraph wires, which carried short and long bursts of current corresponding to the dots and dashes of Morse code. So far as the nature of the message is concerned, they are more like what telegraph wires would be if they could transmit only dots. However complex the information that is to be sent along a nerve may be, it can be sent only as streams of action potentials—action potentials whose size cannot be controlled at the point of stimulation. Information about the most subtle sensations or the most complex actions can be transmitted to and from the brain in no other way.

In 1939, thirty years after Keith Lucas' famous experiment, the poet W.H. Auden, writing about lovers, asked: 'Are her fond responses all-or-none reactions?' The answer to his rhetorical question is that of course the responses are *not* themselves 'all-or-none' in character, but the messages in both sensory and motor nerve fibres that make those responses possible *are* made up of discrete 'all-or-none' units. At a time when our television sets and telephones and cameras are all 'going digital', this feature of nerve messages may not seem much of a limitation, but it is. It is, not directly because of the digital nature of transmission, but because the maximum frequency with which action potentials can pass through a particular point on a nerve fibre is only a few hundred per second—many orders of magnitude less than the frequencies possible in our digital electronic devices. A consequence of this low upper limit is that, though visual information can be transmitted from an aerial to a television set along a single wire, the optic nerves that carry visual information from our eyes to our brains each need to contain about a million individual nerve fibres.

When Bernstein put forward his membrane theory in 1902, it was elegant, it was ingenious, and it was plausible; but there was only a little evidence for it. And that remained the situation until just before the Second World War. Knowing whether Hermann's local-circuit hypothesis was right was important because, if right, it would not only explain why action potentials travel along the nerve, but also remove any lingering doubts about the identity of the action potentials and the message—the fact that the message consists of nothing more than a succession of action potentials. After all, the identity of their speeds along the nerve might be no more significant than the identity of the speeds across the countryside of the railway engine and of the plume of smoke above it. (It is the movement of the engine that causes the movement of the smoke, but they are not the same thing.) In the summer of 1938, Alan Hodgkin,[‡] then a young Cambridge physiologist working in the United States, was arguing about the local-circuit hypothesis with an American colleague, Joe Erlanger. Since part of the hypothetical local current is conducted by the fluid bathing the nerve fibre, altering the resistance of that fluid should, if the local-circuit hypothesis is right, alter the flow of electric current along the outside of the nerve fibre and hence the speed at which the action potential travels along the fibre. Erlanger said he would take the hypothesis seriously only if Hodgkin could demonstrate this effect.

By 1938, the availability of thermionic valves to amplify small changes in voltage, and of cathode ray tubes to measure such changes with virtually no time lag, meant that it was possible to measure the speed of the action potential along the nerve without the paraphernalia used by Bernstein. Using single nerve fibres from crabs, or single giant nerve fibres from squid, Hodgkin showed that if the bathing solution was replaced by oil or air, so that only a thin layer of salt solution was available to conduct current outside the fibre, the speed of the action potential along the nerve was reduced. This was encouraging, but the result was not conclusive because the slowing might have been caused by deleterious effects of the oil or slight deterioration of the fibre when it was suspended in air.

The matter was finally clinched by an experiment that was as simple as it was decisive.[125] Instead of looking at the effect of *increasing* the electrical resistance to the flow of current outside the fibre, Hodgkin decided to look at the effect of *decreasing* it. A giant nerve fibre from a squid was arranged so that the part between the point at which it was to be stimulated and the point

‡ Another member of the Quaker family of Hodgkins.

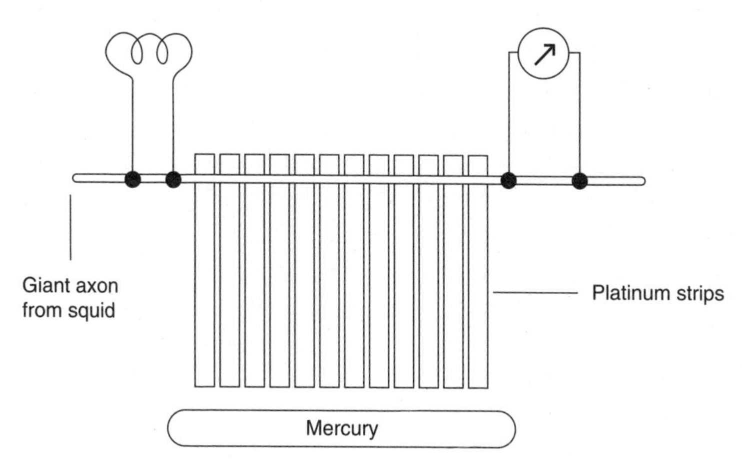

FIG 49 Alan Hodgkin's experiment proving the 'local-circuit hypothesis'. The coil represents a device for stimulating the nerve with a small electric shock, and the meter a device for recording the action potential.

at which the action potential was to be recorded rested on a series of parallel platinum strips, in a moist atmosphere. The strips ended just above a small trough of mercury, and could be connected together electrically by raising that trough—see Figure 49. If the local-circuit hypothesis was right, Hodgkin argued, raising the trough should increase the speed at which an action potential moved along the nerve fibre, and that is what he found. What is more, there was no detectable pause between raising the trough and seeing the change in speed. As nothing in direct contact with the nerve had changed, and the only agent that could have travelled through a metallic short circuit in the time available is an electric current, the experiment provided unequivocal and elegant proof of the local-circuit hypothesis.[§]

[§] It is, incidentally, worth noting that something rather like Hodgkin's elegant method for increasing the speed of conduction has been evolved by natural

In August of the following year, back in England, and working at the Marine Station in Plymouth with Andrew Huxley, Hodgkin did another elegant experiment with the giant nerve fibres of squid, which led to a major modification of Bernstein's membrane theory. That theory, proposed in 1902, assumed that during the action potential the electrical potential difference across the nerve membrane drops to nearly zero; and this was still the accepted view at the beginning of 1939. Although modern techniques made the measurement of rapid changes in the membrane potential easy, it was still necessary to measure the voltage between a point on the outer surface of the nerve fibre and another point on the surface that was damaged so that contact could be made with the fluid inside the fibre. And since the fibre had to be kept moist there was always the problem of short-circuiting through the external fluid. In January 1939, Hodgkin and Huxley realized that, because the giant nerve fibres of squids can be up to a millimetre across, it would be possible to push a fine glass tube (about 0.1 mm in diameter and filled with sea water) along the inside of the nerve fibre, and to measure the potential difference between the tip of the tube inside the fibre and the sea water in contact with the outer surface of the fibre.

selection to increase the speed of conduction in nerves, particularly in vertebrates. In 1878 Louis-Antoine Ranvier described nerve fibres that had a non-conducting sheath of a fatty material called myelin interrupted by a gap every millimetre or so. The result of this, it was later shown, was that instead of local circuits from an active region of nerve exciting the adjacent resting region, they excited the resting region in the next gap. Instead of the relatively slow changes in permeability associated with the action potential having to occur in succession in each bit of nerve membrane, they needed to occur only in the regions exposed in the gaps. In the intervals between the gaps, conduction was simply the result of the local circuits. The progress of the action potential along the nerve fibre was, therefore, by a series of jumps. It turned out that this *saltatory* conduction (from the Latin *salto*, I jump) was about ten times as fast as conduction in a similar size nerve fibre without a myelin sheath.

That would give the magnitude of the resting potential; and the transient change in the potential difference when the nerve was stimulated would give the magnitude of the action potential. They did the experiment and the result was startling.[126] Instead of dropping to near zero, the potential difference actually reversed—the inside of the fibre becoming positive relative to the outside.# Subsequent experiments showed that in the resting nerve the potential difference across the membrane was −70 to −90 millivolts (the negative sign showing that the inside is negatively charged relative to the outside), and at the peak of the action potential the potential difference was about +40 millivolts. The total swing in potential—the action potential—was therefore well over 100 millivolts.

Clearly, the reversal of the potential needed explaining, and an early suggestion was that it might imply a transient and *selective* increase in the permeability of the membrane to sodium ions. The concentration of sodium ions outside nerve fibres is much higher than the concentration inside, and in the resting nerve there is also an electrical potential difference tending to move sodium ions inwards. A sudden increase in the permeability of the membrane to sodium ions would therefore lead to a rapid net entry of positive charge. As the potential difference across the membrane decreased, the near-equilibrium for potassium ions would be disturbed and there would be a net outward movement of potassium ions; but if the permeability for sodium ions were transiently much higher than that for potassium ions, the net inward movement of sodium would be greater than the net outward movement of potassium, and the membrane potential would reverse. Unfortunately, less than three weeks after Hodgkin

Bernstein had sometimes found a reversal of the sign of the membrane potential in his experiments, but because the result was inconstant he ignored it.

and Huxley did their first successful experiment, Hitler invaded Poland and war took over.

It was not until 1945 that Hodgkin and Huxley got back to working on nerves, and in the course of the following two years they made a number of observations supporting the notion that the entry of sodium ions and the exit of potassium ions were connected respectively with the early and late stages of the action potential. But anyone trying to sort out just what happens during the action potential faced two formidable difficulties. Firstly, the observable changes happen very fast—the whole action potential lasting only a few milliseconds. Secondly, the timing of the events at any one point on the nerve will be slightly different from the events at adjacent points—either lagging behind or being ahead, depending on which direction the action potential is moving. Local circuits will therefore complicate observations at any particular point.

In the summer of 1947, K.S. Cole and G. Marmont, working at the Marine Biological Laboratory at Woods Hole on Cape Cod, invented an elegant and, as it turned out, extraordinarily fruitful approach to these problems.[127] Using a long metal electrode that could be put inside a squid giant nerve fibre, and suitable electronic feedback, they arranged to be able to make a sudden change in the voltage across the membrane of a substantial length of nerve, to hold the voltage at the new level, and to record the current flow necessary to maintain that new level. If the voltage was not changing, the recorded current flow at any moment could be assumed to be the same as the net flow of charge through the membrane at that moment. A particularly interesting observation was that a sudden forced 50 millivolt drop in the membrane potential led, among other things, to a transient *inward* current through the membrane—just the opposite of what you might expect from a simple physical system, but just what you would expect if the drop in membrane potential caused a sudden and transient increase in the permeability of the membrane to sodium ions.

A similar technique using an electronic feedback system to provide what would become known as a 'voltage clamp' had been discussed by Hodgkin and Huxley early in 1946, and in 1947 they spent a good deal of time considering the kinds of molecular mechanism that would be required to account for the known features of the action potential. The behaviour of some of these models was readily predicted if the voltage was constant but not otherwise, and this fact, together with the success of Cole and Marmont, encouraged them to embark on a series of 'voltage-clamp' experiments on squid giant nerve fibres, starting in 1948 and finally published in 1952.[128] (This is not as exhausting as it sounds since the squid season at Plymouth lasts only a few months.) They used two wire electrodes wound on a thin glass rod that could be pushed along the inside of a squid giant nerve fibre, one to measure voltage, the other to deliver whatever current was necessary to control the voltage. By varying the size and duration of the voltage changes and noting the effects of varying the composition of the bathing fluid, they were able to follow the changes in the conductivity of the membrane to sodium and potassium ions in a variety of situations.

Figure 50 summarizes some of their most significant results.[129] It shows the changes in the membrane conductivity to sodium and potassium ions following a sudden 56-millivolt reduction in the membrane potential. After a very brief S-shaped start, the conductivity to sodium ions (g_{Na} in the figure) rises steeply to a peak and then quickly falls back

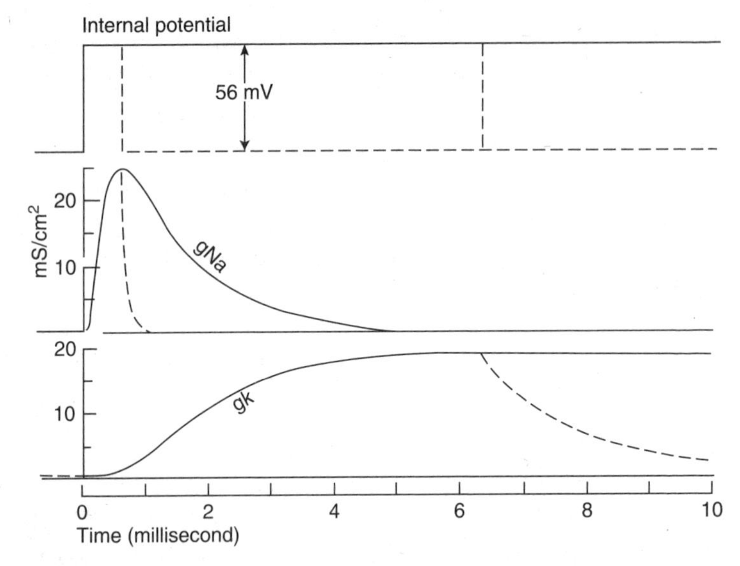

FIG 50 The results of a 'voltage clamp' experiment by Hodgkin and Huxley.

exponentially[##] to its initially low level, *despite the fact that the drop in voltage is maintained*. In contrast, the conductivity to potassium ions (g_K in the figure) rises slowly in an S-shaped curve, gradually reaching a maximum, which is maintained as long as the drop in voltage is maintained. If the voltage is restored to its original level, the conductivity to potassium falls slowly and exponentially—see the dotted line. This difference between the behaviour of the membrane to sodium and potassium ions explains why in a normal (i.e., 'unclamped') nerve a sudden reduction in resting potential leads *firstly* to a rapid entry of sodium ions, reducing the

[##] A fall is said to be exponential if the rate of fall at any moment is proportional to the height at that moment.

voltage to zero and then reversing it, and *secondly* to a rapid loss of potassium ions, restoring the voltage to its original level. At the end of the action potential, the nerve will have gained some sodium ions and lost some potassium ions, and although the changes in concentration are extremely small (except in nerves that are both very small and very active), such ion movements have been confirmed using radioactive isotopes of sodium and potassium. These isotopes make it possible to measure ion movements in each direction—another elegant technique.[130]

In 1963, Hodgkin and Huxley were awarded the Nobel Prize for their work on the mechanism of nerve conduction.[131] During the half-century that has passed since their experiments, much has been learnt about the molecular machinery that accounts for the ion movements during the action potential, and also about the molecular machinery that generates the differences in the concentrations of sodium and potassium on the two sides of the nerve membrane that serve as the immediate source of energy for the transmission of messages along the nerve—and also, incidentally, for several other processes. But I want now to switch from considering the way in which information is transmitted along nerves to the way that information is handled in the brain.

8

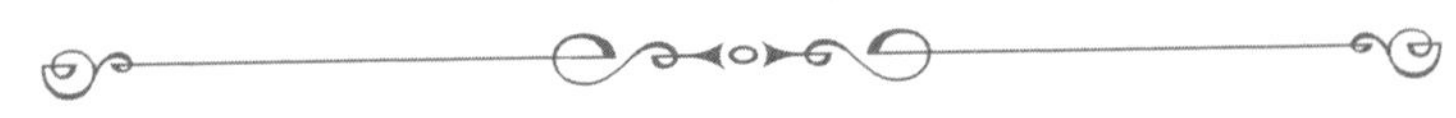

INFORMATION HANDLING IN THE BRAIN

There is a striking difference between the way in which we receive information by post or telephone or the Internet, and the way our brain receives information from our sense organs. The information we derive from a letter or telephone call or email, including the identity of the sender, is all in the message. But, as we saw in the last chapter, the messages our brains receive from our sense organs all consist of very similar impulses—action potentials—and because of the all-or-none law we can derive no useful information from the size of those impulses. It follows that all the information arriving along a given fibre must come from the timing of the impulses or the identification of the fibre. Somehow, these two features must tell us about the site and nature of the stimulus, how intense it is and how long it lasts. You might say that, given the enormous rate at which information nowadays is transmitted in digital signals, this limitation is not serious; but you would be wrong for two reasons. Firstly, as we have seen, nerves cannot transmit impulses at frequencies of more than a few hundred per second. And secondly, even at these low frequencies, encoding information in intricate timing patterns is impossible because a nerve fibre that has just conducted several impulses in quick succession will conduct more slowly, so the pattern will become distorted as the message proceeds along the nerve.

In fact, the frequency of impulses in sensory nerves usually reflects the *intensity* of the stimulus, and all other information must come from the identification of the particular fibre. That is why, in any small area of skin, there will be receptors for touch, for cold, and for heat, each with its own nerve fibre to the brain or spinal cord: in the cochlea of the ear there will be nerve fibres from each part of the 'organ of Corti'—the long membrane whose different parts resonate with sounds of different pitch; in the retina there will be cells sensitive to light of different colours; and so on. The way the brain makes use of this enormous inflow of information to build up a picture of current events, and to initiate appropriate responses to those events, is a huge subject, and one in which there has been great progress in the last two centuries. A vast amount has been learnt from experiments on animals, from experiments on normal people, from investigations of patients with damaged brains, and more recently from sophisticated scanning techniques and from advances in our understanding of information processing. All I want to do in this chapter, though, is to illustrate elegance, firstly by talking about a few surprising experiments that throw some light on the way we analyse information from our eyes; secondly by describing some elegant analyses of the bizarre disabilities of three patients with damaged brains; and thirdly by showing how a curious and important feature of colour vision, already noticed in the eighteenth century, was explained in the late twentieth century by a combination of ingenious theory and the counterintuitive results of experiments.[132]

In the third quarter of the nineteenth century, Helmholtz pointed out that, even in forming perceptions about straightforward physical features in the world about us from the information we get from our eyes, we must rely to a considerable extent on unconscious inductive inference—probably inference from experiences during our early childhood. He argued like this: everyone would accept that someone wandering about a familiar but ill-lit room, with objects scarcely discernible, is able to find his way without disaster by making use of knowledge from earlier visual impressions. But even when we wander round a room flooded with sunshine, 'a large part of our perceptual image may be due to factors of memory and experience...Looking at the room with one eye shut, we think we see it just as distinctly and definitely as with both eyes. And yet we should get exactly the same view if every point in the room were shifted arbitrarily to a different distance from the eye, provided they all remain on the same line of sight.'[133] The drawing by Oliver Braddick in Figure 51 shows why.

It follows that if we keep still and use only one eye, the images on our retinas are compatible with a host of differently shaped objects. What we see, Helmholtz pointed out, is the object that is *usually* responsible for forming that image. From where I am writing I can see a plate hanging on the wall some distance to my right. It is a round plate and I see it as a round plate, but the image on my retina must be oval, and an oval plate hung at an appropriate (but highly unusual) angle to the wall would give the same image. Helmholtz points out that a blow on the outer corner of the eye gives the impression of light somewhere in the direction of the bridge of the nose, because 'If the region of the retina in the outer corner of the eye is to be stimulated [by light], the light has to enter the eye from the direction of the bridge of the nose.'

A particularly elegant example of this kind of argument was demonstrated to me by the psychologist Richard Gregory. I had

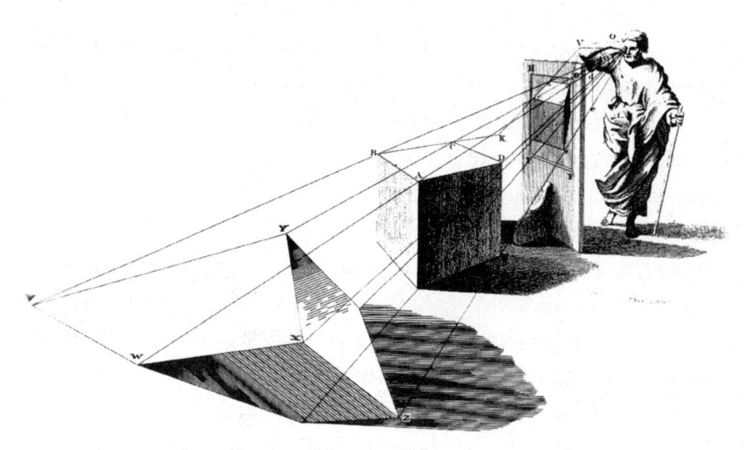

FIG 51 Drawing to illustrate Helmholtz's argument. To the venerable figure looking with only one eye, the curiously shaped object in the bottom of the picture is indistinguishable from a cube, because all the identifiable features are on the same lines of sight as they would be for a cube.

to sit in a dimly lit room, staring at a large white cube. He stood behind me, and when I had become accustomed to the dim light he fired a flashgun illuminating, momentarily but intensely, what I was looking at. After the flash, I felt that I was still seeing the cube, but the activity in my retina was the aftermath of the vigorous activity that had occurred during the flash; what I was really seeing was an after-image. He asked me to continue looking at the cube while slowly rocking forwards and backwards; to my surprise the size of the image did not change, and what I seemed to see was the cube receding and advancing, but always keeping the same distance from my face—an illusory change that precisely fits Helmholtz's statement.

And Gregory had another trick. In the next experiment I had to stare at my own hand held out in front of me; and after the flash, instead of rocking forwards and backwards, I had to move my hand alternately towards and away from my face. Again the size of the image remained constant, but because information from my muscles and joints made me aware of the real movements of my hand, the explanation of the unexpected constancy that served to explain the apparent behaviour of the cube in the first experiment was not available. What I seemed to see was that as I moved my hand towards my face it shrank alarmingly; as I moved it away it grew. Again this fits with Helmholtz's statement, since it is only by shrinking and growing that an object moving alternately towards and away from the face can keep the size of its retinal image constant.

An interesting feature of these experiments is that, as in Helmholtz's blow to the outer corner of the eye, the subject's illusory sensation is immediate and effortless. There is no conscious weighing up of alternative hypotheses. It is only after the illusion has been experienced that the subject, if so inclined, can try to rationalize it.

Sometimes our perception is based on unconscious inference from a fact that we have never thought about, and may even have been unaware of. Figure 52 is a picture produced by Vilayanur Ramachandran, which seems to show a series of smooth bumps on the left and a series of smooth hollows on the right.[134] But if you rotate the picture through 180°, the bumps still seem to be on the left and the hollows on the right. His explanation is that the inference of shape from shading is ambiguous, depending on the direction of the light, and we tend to assume that light comes from above because in the world we live in objects tend to be illuminated from above rather than from below.

I want now to turn to what I earlier referred to as the elegant analyses of the problems of patients with damaged brains. Elegant may

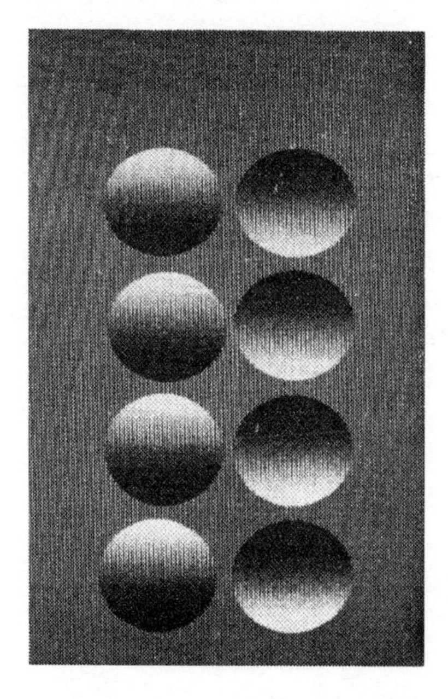

FIG 52 Ramachandran's bumps and hollows.

seem an odd word in this context. One normally thinks of any analysis of a patient's disorders as sound or unsound, plausible or implausible, helpful or unhelpful. But the three very different patients I want to discuss each presented such a bizarre collection of signs and symptoms that at first sight each case seemed incomprehensible and likely to remain so. In that situation an analysis leading to an explanation that is concise, surprisingly simple, highly satisfying,* and probably correct, deserves the accolade of 'elegant'.

In 1988, a young Scotswoman—a rather enterprising Scotswoman with a private pilot's license—was working in northern

* The satisfaction, alas, was restricted to the investigators and those who read about the explanation later. In none of the three cases did the explanation lead to a cure or even an alleviation of the condition, and in two of the cases it was a post-mortem examination that finally showed that the explanation was almost certainly correct.

Italy as a freelance commercial translator. One day she took a shower at her home and was poisoned by carbon monoxide from a faulty water heater. She—I'll call her by her initials, DF—was admitted to hospital in a deep coma, and when she recovered consciousness was found to be blind. And though, after about ten days her vision gradually returned, it was far from normal. A year later, she still couldn't recognize objects that were put in front of her. When she was shown a drawing of an apple or an open book, she had no idea what they were, nor could she copy them. It wasn't that she couldn't control her pencil or that she had forgotten what things looked like, for she could draw an apple or a book *from memory* quite well, though she would later be unable to recognize her own drawings.

Fifteen months after her accident, her strange disability was investigated by Melvyn Goodale, David Milner, and their colleagues, in Ontario.[135] They sat her in front of an upright disc with a large slot in it whose orientation could be made vertical or horizontal or any angle in between by rotating the disc. They then put a card in her hand and asked her to turn it so that its orientation matched that of the slot. She could not manage this, sometimes even failing to distinguish between vertical and horizontal. On the other hand, when she was asked to 'reach out and post' the card through the slot, she behaved just like a normal subject, adjusting the orientation of the card appropriately, long before it was anywhere near the slot.

Since posting a card through a slot is a more familiar activity than turning a card to match the slope of a slot, you may wonder whether she simply failed to understand what she was being asked to do. But that's unlikely, because when she was asked to shut her eyes and to turn the card in her hand to match the slope of an *imaginary*, instead of a real, slot at 0°, 45°, or 90° from the vertical, she could do that perfectly well. So what does this tell us? *Well, her ability to post the card without difficulty shows that information about the orientation of the real slot must have reached her brain and been available for the direct visual control of movement. But it was not, apparently, available to her for forming a conscious perception.*

The question then arose as to whether a similar pattern applied to other aspects of visual form. Normal people are very good at discriminating between shapes—a slight elongation of two opposite sides of a square, for example, turns it into an obvious oblong—and a standard test used by neurologists is to see how good patients are at discriminating between a series of rectangular plaques of equal area ranging from a square to an oblong twice as long as it is broad. Goodale and his colleagues prepared two sets of such plaques and presented them to DF, two at a time in all possible combinations, asking her simply to say whether the two were alike or not. In this she scored no better than chance. And when she was shown the plaques one at a time and asked to use the index finger and thumb of her right hand to indicate the width of the plaque in front of her she again failed. There was no correlation between the width of the plaque and the distance between her finger and thumb. Yet when they asked her to reach out and pick up a plaque she behaved just like a normal subject. Again you may wonder whether she simply didn't understand what she was being asked to do, but that won't wash because when she was asked to close her eyes and show with her finger and thumb the width of an *imaginary* plaque of given dimensions, she did that perfectly well. So again we have to conclude that *visual information that was available to her for guiding a grasping movement was not available to her for answering simple questions about size and shape*. Incidentally, she had no difficulty in identifying colours or surface textures.

So far, I have talked almost exclusively about DF's behaviour during experiments, but the same peculiarities were apparent in her ordinary life. Standing in front of a door, she couldn't describe the shape or position of the doorhandle, but she could reach out and grasp it and open the door just like anyone else.

What makes this pattern of behaviour so surprising is that we tend to take it for granted that in performing a deliberate conscious act like posting a card through an unfamiliar slot, or opening an unfamiliar door, we first form a conscious perception of the shape and position of the slot or the handle and then adjust our movements to suit—in other words our conscious mental processes control our movements. Clearly, this common-sense view doesn't fit the behaviour of this unfortunate woman.

The suggested explanation of her strange behaviour, which is almost certainly correct, is that the carbon monoxide poisoning damaged the machinery necessary for having conscious perceptions of shape and position and orientation, but *not* the machinery for making use of information of that kind to control movements, and also, of course, *not* the machinery necessary for having conscious perceptions of colour or surface texture, or for answering questions about these perceptions. Since, in the course of evolution, mechanisms for the visual control of movement arose long before mechanisms for thinking about formal geometrical questions, considerable separation of the neural pathways involved is just what one would expect.

The next patient whose problems I want to discuss—a Monsieur C—was an intelligent and musical Parisian fabric designer and merchant, in the late 1880s and early 1890s. His problems were primarily with reading and writing, and as both reading and writing depend on our ability to speak and to see, I need to begin

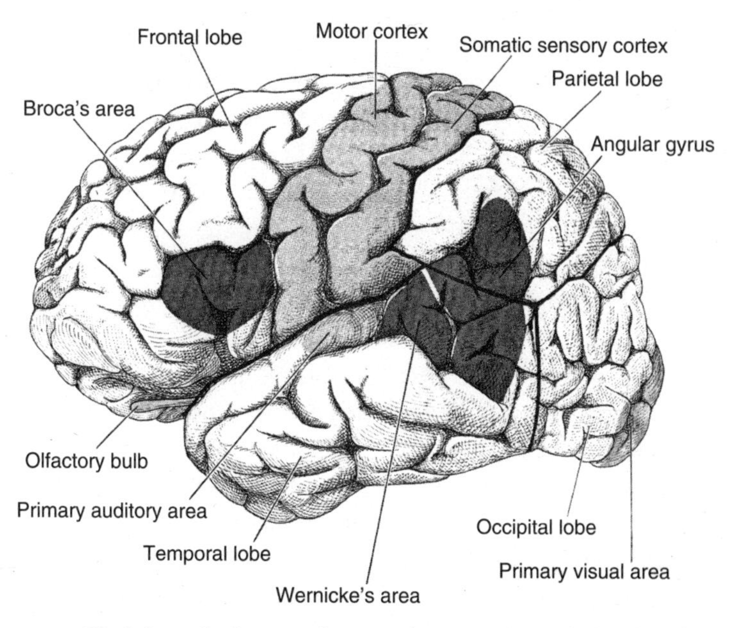

FIG 53 The left cerebral cortex showing the areas particularly relevant to speech.

by summarizing what was known at that time of the role of different parts of the brain in those tasks.

In the 1860s and 70s studies of the correlations between speech difficulties and brain damage (seen post mortem), had shown that two areas of the cerebral cortex, *both on the left side of the brain*, were vital for speech[136]—see Figure 53. 'Broca's area', discovered by Paul Broca in Paris, was in the left frontal cortex adjacent to areas controlling muscle activity of the mouth and tongue, and damage to it was correlated with difficulty in finding and forming words, giving a sparse telegram-like style of speaking but no loss of comprehension. 'Wernicke's area', discovered by Carl Wernicke in Breslau, was in the left temporal lobe adjacent to the area concerned with hearing, and damage to it was correlated with

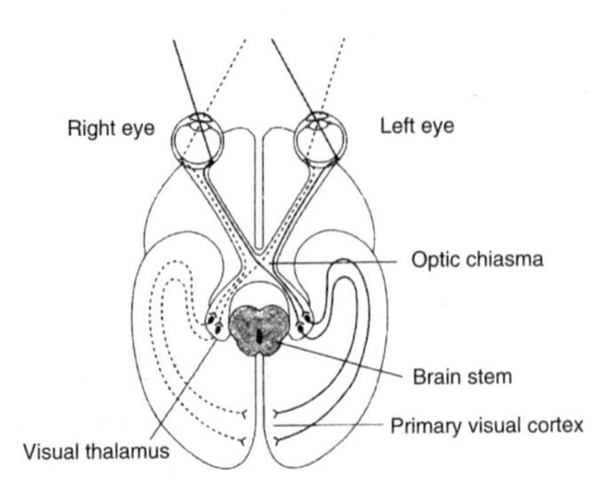

FIG 54 The optic pathways. The brain is viewed from below.

loss of comprehension, and a fluent speech sounding superficially normal but lacking in content and with many wrong words, malformed words, and invented words. It seemed likely, then, that Broca's area contained neural machinery involved in forming spoken words, and Wernicke's area contained neural machinery involved in recognizing and understanding spoken words.

Turning to 'seeing', at Uppsala in the late 1880s, Saloman Henschen showed that there was a strong correlation between loss of vision in parts of the visual field and damage to the cerebral cortex in the hindmost part of the occipital lobes of both hemispheres. Loss of this part of the occipital cortex (later referred to as the primary visual cortex) on one side was always associated with blindness in the *opposite* half of the visual field in *both* eyes. This apparently odd finding makes sense. Herman Munk, who had earlier obtained similar results in monkeys, showed that the optic nerve fibres from the nasal half of each eye cross over—see Figure 54—with the result that all information about the left half of the visual field is conveyed to the primary visual cortex of the right occipital lobe, and all information about the right half of the visual field is conveyed to the primary visual cortex of the left occipital lobe. The need for this curious arrangement is simply that, in both eyes, the nasal and lateral halves of each retina receive information from different halves of the visual field. The crossing over makes it easy for the brain to compare the view of each part of a scene as it appears to the right and left eye, and it is the difference between those views that makes stereoscopic vision possible.** The two cerebral hemispheres are connected by a tract—the *corpus callosum*—containing more than 200 million nerve fibres, so that normally information can be exchanged easily between the two hemispheres. It is a measure of the efficacy of the corpus callosum, that though all the information from each half of our visual field is fed initially only to the occipital lobe on the opposite side, we are never conscious of a seam down the middle of the field.

Now to return to Monsieur C, the Parisian fabric designer. In the autumn of 1887, he had a number of transient attacks of numbness and slight weakness in the right arm and leg, together with a slight difficulty in speaking. Although these attacks continued for several days, he was able to take long walks about Paris during which he had no difficulty in reading the shop signs and posters. On the sixth day, though, he was shocked to find that he could not read a single word; yet his recognition of objects and people seemed entirely normal, he had no difficulty in speaking, and-despite his inability to read—no difficulty in writing. His vision had previously been particularly good, and in designing fabrics he had had no difficulty in counting threads and working out patterns on paper marked in millimetre squares. Assuming that something must have gone wrong with his eyes, he consulted an ophthalmologist who referred him to the physician Jules Dejerine.[137]

Both the ophthalmologist and Dejerine noted that, as well as suffering from an inability to read—what neurologists call

** If, sitting in a room and keeping your head still, you shut one eye, the loss of stereoscopic vision is dramatic.

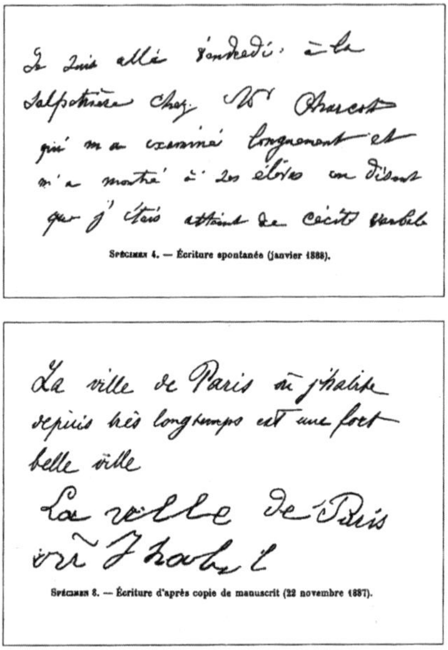

Spécimen 4. — Écriture spontanée (janvier 1888).

Spécimen 8. — Écriture d'après copie de manuscrit (22 novembre 1887).

FIG 55 The handwriting of Dejerine's patient, Monsieur C.

alexia—Monsieur C was blind in most of the *right* half of the visual field of each eye, though vision in the left half of the field was normal. This selective loss of vision suggested severe damage to the visual cortex in the *left* occipital lobe.

Although he couldn't recognize words or letters, Monsieur C could still write, either spontaneously or to dictation, almost as well as before the attack—see Figure 55 upper box. But he couldn't read what he had written unless the writing was so large that he could follow the shapes of the individual letters with his finger. He also found it very difficult to copy either print or cursive writing, needing to refer constantly to the text he was copying, and proceeding slowly and laboriously as if the letters were unfamiliar designs (Figure 55 lower box). He completely lost his ability to read music, though he retained his interest in operas and could still sing arias and even, with his wife's help, learn new ones. Though he could not read letters or music, he was still able to read numbers and he could do arithmetic, if somewhat untidily; astonishingly, he negotiated an annuity for a friend so skilfully that the insurance office assumed he was a professional agent and offered him a commission. He was also able to recognize *Le Matin*—his regular newspaper—by its heading, though not other newspapers, suggesting that he could recognize some familiar patterns; and this fits with his continued ability to play cards.

For the next four years, Monsieur C's medical problems remained much the same. With his wife's help, he led a more-or-less normal life, though with occasional lapses into depression, once even considering throwing himself from the column in the Place de Vendôme. In January 1892, he had a slight stroke; he remained conscious but there was weakness on the right side of his body, and his speech became incomprehensible, with wrong words and non-words. By the next day the weakness had gone, but as well as being unable to communicate by speaking he now found that he could not write, and was reduced to making signs. Yet his intelligence and ability to understand speech seemed to be unaffected, and he remained in this state until he became comatose and died ten days later.

It was the autopsy that finally revealed the causes of Monsieur C's disabilities. As expected, his brain showed two distinct kinds of lesion: old yellow lesions presumably related to the features that had persisted over four years, and fresh lesions presumably related to the features that followed the final attack. Also as expected, the old lesions had totally destroyed the visual cortex of the left hemisphere; but they had in addition damaged some nerve fibres from the hindmost part of the corpus callosum—the part responsible for exchanging *visual* information between the two hemispheres. With the left visual cortex destroyed, letters and words could have been seen only through signals reaching the right visual cortex, but for these letters and words to have been recognized, information from the right visual cortex would have had to be transferred through the corpus callosum to speech areas in the left hemisphere. Damage to the hindmost part of the corpus callosum could therefore account for Monsieur C's inability to read or to copy writing. (His ability to recognize figures suggests that fibres from the corpus callosum connected to the part of the cortex involved in figure recognition had not been destroyed.) So much for the old lesions.

The new lesions were restricted to the region of the left angular gyrus, which lies just behind Wernicke's area and is thought to link parts of the cortex concerned with hearing and with vision. A year before Dejerine was consulted by Monsieur C, he had had a patient who had suddenly become unable to read or to write, and whose speech often contained wrong or malformed words. About nine months later this patient died, and at autopsy his brain showed damage restricted to the region of the angular gyrus. Dejerine therefore thought it likely that damage to the angular gyrus prevented the two-way interconversion of the representation of the sound of a word and the representation of its appearance. It follows that damage to the angular gyrus would be likely to interfere with both reading and writing. In Monsieur C's case the old yellow lesions already accounted for his inability to read; the new lesion affecting the left angular gyrus would account for his inability to write, and also for his deteriorating speech.

And there is an interesting tailpiece to the story. If Dejerine's hypothesis to account for Monsieur C's inability to read is correct, then in a patient with a healthy visual cortex on both sides, you would expect that damage to the rear part of the corpus callosum would cause reading problems in the left half of the visual field but not in the right half.[†] In other words, the patient looking straight forwards would be able to read something to the right of the midline in the field of view but unable to read the identical text placed to the left of the midline. Twenty years after Dejerine's death, and nearly fifty years after he was consulted by Monsieur C, precisely this picture was found at Johns Hopkins Hospital, Baltimore, in a young woman whose corpus callosum had been partly cut through so that a small tumour could be removed.[138]

The last case I want to discuss is another case of carbon monoxide poisoning but with consequences very different from those suffered by DF, and even more tragic. It was described by Norman Geschwind, Fred Quadfasel, and José Segarra, in Boston in 1968.[139] The patient, in the care of Quadfasel, was a 21-year-old woman who had been found in her kitchen unconscious and not breathing, with the unlit gas jet of the water heater turned on. Nearly a month passed before she regained consciousness, and when she did she was confused, incontinent, and with almost completely paralysed limbs—a state she remained in for nearly ten years. She could recognize people and she smacked her lips or made repetitive movements of the hands to indicate satisfaction or dissatisfaction. Except for a few phrases—'Hi, Daddy!' 'So can Daddy,' 'Mother,' 'Dirty Bastard'—she never spoke spontaneously, and she did not seem to understand what was said to her. Her response to questions or phrases was to repeat the question or phrase, which she did in a normal voice and with good articulation. This repetition—*echolalia*, the neurologists call it—was very marked, but occasionally, when the phrase was a familiar one,

† Not in the right half, because information from the right half of the visual field is handled by the visual cortex of the left occipital lobe, which has access to the speech areas without involving the corpus callosum.

instead of repeating it she might complete it. Hearing 'Ask me no questions,' she might say, 'Tell me no lies.' Told 'Close your eyes,' she might say, 'Go to sleep.' Occasionally she would pick up a word used by the examiner and let it trigger a conventional response. For example, if asked: 'Is this a rose?' she might reply, 'Roses are red, violets are blue, sugar is sweet, and so are you.'

Features of this kind had been reported in patients before, but Quadfasel's patient showed something new. In a room with a radio playing songs, she tended to sing along with the radio. If the radio was turned off and the song was familiar, she would continue to sing both words and tune correctly for a few lines. Similarly, listening to a religious broadcast, she would recite along with the priest. More remarkably, if she was exposed repeatedly to an unfamiliar song she could learn both the tune and the words.

During the nine years she was looked after, and carefully observed, she was never heard to make a spontaneous statement or a request. And, apart from the peculiar responses already described, she never gave the impression that she understood what was said to her. It looked as though the machinery for hearing speech, the machinery for producing speech, and the machinery for remembering speech were all working, but working in isolation from other parts of the brain.

And that picture was confirmed when, nearly ten years after the original poisoning, Quadfasel's patient died. A detailed post-mortem examination of her brain showed very large areas of destruction, but Broca's area, Wernicke's area, and the pathway connecting them, had survived, as had the auditory pathways, the parts of the motor cortex involved in speech, and, deep within the forebrain, most of the hippocampal regions—regions known to be necessary for the formation of memories. In other words, the areas needed for producing speech, for hearing speech, and for remembering speech were all intact, but isolated from other parts of the brain. The pathological lesions were wholly consistent with the elegant, though depressing, analysis of the situation reached by Geschwind and his colleagues from their observations of the patient's behaviour.

I want to finish this chapter by considering the role of the brain in colour vision, and in particular its role in accounting for a feature to which attention was first drawn in the late eighteenth century. The feature—generally known as 'colour constancy'—is simply that objects usually seem to have much the same colour even if viewed under very different illumination. We don't notice a great change between the way colours look under a cloudy sky or a blue sky, in the morning or at sunset, by electric light or candle light, though it's true we are wary of choosing clothes under fluorescent light. This ability to 'discount the illuminant'—to use Helmholtz's phrase—is obviously useful, since we live in a world in which lighting changes greatly and it must be easier to identify objects if their colours scarcely alter as the light alters. But how is this constancy achieved?

Before considering that question, I need to say something about the way the eyes discriminate between different colours. At the very beginning of the nineteenth century, Thomas Young, who had just produced powerful evidence that light consisted of waves rather than particles (see Chapter 6), pointed out that it was inconceivable that each small area of the human retina had separate receptors for each of the vast number of colours that we can identify. He suggested that each area of the retina need contain only three kinds of receptor, one kind stimulated best by the shorter-wavelength light near the violet end of the spectrum, a second kind stimulated best by the medium-wavelength light in the middle of the spectrum, and a third kind stimulated best by the longer-wavelength light towards the red end of the spectrum. Indeed, he suspected that his contemporary, the chemist John Dalton—who said that he could detect only two hues in the solar spectrum, one corresponding to the normal observer's red, orange, yellow, and green, and the other corresponding to blue and violet—lacked the long-wavelength receptor. Dalton himself thought that the fluid in his eyes was probably blue, and absorbed the red light before it could reach the retina. He asked that his eyes should be kept after his death to test his theory, but the fluid turned out to be colourless. So Dalton was wrong, but so too, as it later turned out, was Young. Astonishingly, in the 1990s, the remains of Dalton's eyes were still in the care of the Manchester Literary and Philosophical Society, and a quartet of scientists from London and Cambridge examined the surviving DNA and found that it was the middle-wavelength receptor that was missing.[140]

Young's 'trichromatic theory' remained largely buried in the Royal Society's *Philosophical Transactions* until it was taken up by Helmholtz in 1850, and it has flourished ever since, explaining not only the existence of various forms of colour blindness but also many of the responses of the normal retina. When an isolated area of uniform colour is seen against a dark background, its perceived colour is determined solely by the balance of wavelengths in the light reflected from it—a balance that controls the relative strength of stimulation of the three kinds of retinal receptor. These receptors were later identified as cone-shaped cells, and in the 1960s it was shown (as had long been believed) that there were three kinds of cone with different light-absorbing pigments, whose absorption spectra fitted their different sensitivities to light of different wavelengths. The theory explained the results of mixing light of different colours and pigments of different colours‡ but it didn't explain colour constancy. If you change the balance of wavelengths in the illuminating light, you will change the balance of wavelengths in the light reflected from whatever you are looking at, so its colour ought to change. Why then do changes in illumination often have so little effect on perceived colour?

Helmholtz felt that the explanation was a process of 'discounting the illuminant' by 'unconscious inference'. He wrote:

> By seeing objects of the same colour...in spite of the difference of illumination, we learn to form a correct idea of the colour of bodies, that is to judge how such a body would look in white light; and since we are only interested in the colour that the body retains permanently, we are not conscious at all of the separate sensations which contribute to form our judgement [since the determination of colour is]...not due to an act of sensation but to an act of judgement.[141]

Which is mostly true but does not get us very far.

The modern solution to the problem of colour constancy began with the observations of Edwin Land, the inventor of Polaroid film. In the 1950s he began a series of remarkable experiments, which showed that it is only in rather narrowly defined circum-

‡ Of course, mixing light of different colours and mixing pigments of different colours give different results. If you look at a piece of white paper illuminated by a mixture of green light and red light, your eye will receive both. The green light will stimulate your middle-wavelength receptors strongly and your long-wavelength receptors less strongly; the red light will stimulate your long-wavelength receptors strongly and your middle-wavelength receptors less strongly. With the intensities of the red and green lights suitably adjusted, the result will be indistinguishable from the result obtained by illuminating the paper with yellow light, and the paper will look yellow. Mixing red and green *pigments* will produce quite a different effect, since the red pigment will absorb most of the middle- and short-wavelength light, and the green pigment will absorb most of the long- and short-wavelength light. The result is that little light is reflected and the paper will look brown.

stances that the colour seen by an observer looking at a particular small area is determined solely by the relative intensities of light of different wavelengths coming from that area. Where the area is part of a scene, the apparent colour depends also on the light coming from the rest of the scene. Two of Land's experiments are worth discussing in some detail, as they show how relatively simple procedures can produce results that are as informative as they are counter-intuitive.

In the first experiment,[142] he used a specially designed camera to take two simultaneous black and white photographs of a young woman, one taken using a red filter (so that the film recorded only long-wavelength light), and the other taken using a green filter (so that the film recorded mostly middle-wavelength and some short-wavelength light). The photographs were positive transparencies, so they could be projected onto a screen.

When the two photographs were projected separately, using white light, the results were exactly as you might expect. For example, though both appeared on the screen as black and white photographs, in the one taken using a red filter (a long-wavelength photograph) the woman's lips and red coat appeared light grey, whereas in the one taken using a green filter (a middle- and short-wavelength photograph) the lips and coat appeared black. When the long-wavelength photograph was projected alone, but through a red filter, the image on the screen was, as expected, in various shades of red grading into black.

The surprise came when the image of the long-wavelength-photograph *projected through a red filter* and the image of the middle- and short-wavelength-photograph *projected without any filter* were exactly superimposed on the screen. Since the screen was receiving red light from one projector and white light from the other, and both transparencies were black and white, common sense would suggest that the result would have been a picture of the woman, in colours ranging from red through pink to white. And straightforward application of Young's trichromatic theory would lead to the same prediction. In fact, what appeared on the screen looked like a full-colour portrait—'blonde hair, pale blue eyes, red coat, blue-green collar, and strikingly natural flesh tones.' For some unknown reason, the classical rules were not working. What was even more surprising, Land found that substantial changes in the relative strengths of the two projectors had little effect on the colours seen.

In the second experiment,[143] Land studied the effects of changing the illumination on the appearance of abstract collages of rectangular coloured papers of various sizes—nicknamed colour Mondrians because of their resemblance to paintings by the Dutch painter Piet Mondrian. He arranged to illuminate a colour Mondrian simultaneously with three projectors, one using long-wavelength (reddish) light, one using middle-wavelength (greenish) light, and one using short-wavelength (bluish) light. The intensity of light from each projector could be individually controlled, and the intensity of the light of each kind reflected from any particular rectangle could be measured with a photometer.

As expected, with the three projectors shining light of similar intensity, the colours of the individual rectangles looked normal, i.e., the same as in white light. More surprising was what happened—or rather what didn't happen—when Land varied the intensities of the light from the individual projectors, though always ensuring that all three were contributing some light and that the sum of the three intensities remained constant. Substantial changes in the relative intensities had little effect on the perceived colours of the collection of rectangles.

With a certain setting of the three projectors, Land first measured the amounts of reddish, greenish, and bluish light reflected by a particular yellow rectangle, and calculated their ratios. He then turned his attention to a particular green rectangle, and adjusted the relative strengths of the three projectors so that the amounts of the three kinds of light reflected from the green rectangle were in the *same* ratios as they had been in the light reflected from the yellow rectangle with the original setting of the three projectors. Astonishingly, when he looked at the whole Mondrian, the green rectangle still looked green. (And, despite the changes in the relative strengths of the three projectors, the yellow rectangle still looked yellow). This was unequivocal proof that the ratios of the different wavelengths reflected from a small area of uniform colour forming part of a multicoloured scene cannot alone determine the colour of that area as seen by an observer. Such behaviour, he emphasized, was quite different from that seen if an *isolated* patch of colour is viewed on its own against a dark background, when its perceived colour *is* determined solely by the ratio of the light of different wavelengths reflected from it. It seems, then, that when we view the whole scene (the assembly of coloured areas), we must alter our interpretation of the information coming from each area by making use of information from outside that area.[144]

How might this be done? The relative degree of stimulation of the three types of cone in any tiny spot in the retina tells us the ratios of the intensities of long-, middle-, and short-wavelength light reaching us from the bit of the scene whose image lies on that spot. Land's suggestion was that instead of judging the colour of any particular bit of the visual scene directly from the ratios of the intensities of light of long, middle, and short wavelengths reaching us from it, we first adjust those intensities to take account of the properties of the rest of the scene. This might be done most plausibly by dividing the absolute intensity of light in each waveband reaching us from the particular area by the *average* intensity of light in that waveband reaching us from *the whole scene.*[145]

This sort of adjustment would explain why changing the relative strengths of the two projectors in Land's first experiment had little effect on the colours perceived in the image of the young woman. But why did those colours appear so magically in the first place? How could blue eyes and a blue-green colour be produced by superimposing a black and white picture and a black and red picture? That too follows from Edwin Land's theory, though the explanation is less obvious.

Consider the projected image of the blue-green collar. The light falling on that part of the screen was mostly white light with a trace of red light; on Thomas Young's theory, the image of the collar should have looked a pale pink.

But now remember that we are not working according to the gospel of St Thomas but the gospel of St Edwin. What determines the perceived colour of a particular area is not the ratios of the absolute intensities of light of long, middle, and short wavelengths reaching us from that area, but the ratios of those intensities *each expressed as a fraction of the average intensity of light of that wavelength reaching us from the whole scene*. When the white projector alone was turned on, the image of the collar on the screen looked pale grey. When the red projector was turned on as well, very little red light reached the screen image of the collar, but a good deal reached other parts of the picture—the lips, the red coat—so the *average intensity of long-wavelength light reflected from the whole picture was much increased*. The intensity of the long-wavelength light reflected from the image of the collar expressed as a fraction of that average intensity was therefore much reduced. Since what determines the perceived colour is not the ratios of the absolute intensities of light in the three wavebands, but the ratios of those intensities each expressed as a fraction of the average intensity of that waveband over the whole picture, the collar looked blue-green. It is as though shining red light on part of the screen subtracts red light from other parts. Removing red light from white light leaves blue-green light.

Land's work provided elegant proof of an elegant theory to explain colour constancy, but it left open the question: where is the neural machinery that does the necessary calculations and how does it work? It must be beyond the primary visual cortex, because a feature of all cells in the primary visual cortex whose responses have been recorded is that in each case the response is wholly determined by the pattern of information in a very small area of retina; the rest of the scene is irrelevant. Since it is perceived colours rather than patterns of wavelengths that we recognize, there ought also to be nerve cells somewhere in the brain whose responses correlate strictly with the colour *perceived* in a particular part of the visual field *irrespective of the particular pattern of light on the retina that gives rise to that colour.* And it seems that there are.

In 1983, in an experiment on an anaesthetized monkey, Semir Zeki recorded the activity of cells in an area of the cerebral cortex (V4) known to be concerned with colour, while the monkey viewed a colour Mondrian.[146] The Mondrian could be moved so that the receptive field of the cell under investigation received light from only one rectangle at a time, and the proportion of light of different wavelengths falling on that rectangle could be adjusted to any desired value. Zeki found that the cells fell into two classes. For those in the first class, the cell's response seemed to depend only on the proportion of light of different wavelengths reflected from the rectangle, and, therefore, illuminating the cell's receptive field. Changing the illumination outside the cell's receptive field had no effect. For those in the second class, firing was always correlated with the colour of the rectangle as perceived by Professor Zeki and his colleagues looking at the whole

Mondrian—a colour that was affected by changes in illumination of the retina outside the cell's receptive field. They could not of course know what the monkey's perceptions would have been had it not been anaesthetized, but it seems reasonable to suppose that the monkey would have been able to make similar discriminations. Zeki's experiment shows where the relevant machinery is, and Land's experiments tell us roughly what it does. Precisely how it does it remains to be discovered; but that doesn't detract from the elegance of Land's elucidation of colour constancy, or of Zeki's demonstration of the existence of cells that behave just as Land's hypothesis predicted some cells should.

9

THE GENETIC CODE

In the first chapter of his book on the human genome, Matt Ridley describes the discovery of the structure of DNA (deoxyribonucleic acid) by Crick and Watson as uncovering 'the greatest, simplest, and most surprising secret in the universe'.[147] This may of course be an overstatement, but his enthusiasm is understandable. Crick and Watson's work, and the long sequence of work by others that led up to it, was remarkable in various ways—the range and elegance of the investigative methods used; the fact that (quite unexpectedly) knowing the structure made it pretty clear how DNA does its job; the fact that that job is to enable living things to replicate, and to replicate in such a way that evolution by natural selection is possible; and, above all, the extraordinary accuracy, reliability, and indeed elegance of the machinery that their work revealed (though scientists cannot, of course, claim any credit for that). Here, then, we have the happy combination of elegant investigations leading to important and elegant results. As usual, the full story is more complicated and has interesting roots and consequences.

The long history of work that culminated in the Watson–Crick model, and what it implies, can best be thought of as a 'play in four acts', extending over a century but with a surprisingly long interval between the first and second acts. Each of the acts describes one of the four crucial steps that together led to our present understanding. The first step was quite unexpected. It was obvious that for each living organism to produce appropriate offspring, a very large amount of information must be transferred from one generation to the next, and that information must somehow be contained in the germ cells that form the offspring; yet how the information was stored, and how it was used, remained a complete mystery. In the 1850s and 1860s, however, detailed studies of plant hybrids, and of the offspring of hybrids, showed that the inheritance of individual characters often follows rather simple patterns, and those patterns give important clues to the way the information is packeted and handled. The second step was the realization that complete sets of the packets of information—we now call such packets *genes*—are held in the nuclei of nearly all living cells,[148] in microscopic structures we now call chromosomes. The third step was the identification of the kinds of molecule inside chromosomes that encode the information. And the fourth was the discovery of both the way in which the information is encoded in those molecules, and the way in which that information is used to synthesize the proteins necessary for constructing the offspring. The four steps involved very different people, taking very different approaches and using very different techniques, but they all involve elegant theories or elegant experiments or both. They are worth looking at in more detail.

ACT 1
THE TRANSFER OF INFORMATION FROM GENERATION TO GENERATION

A century before Crick and Watson's discovery, Darwin was struggling with the concept of evolution through natural selection—the theory that the selective survival of the fittest among a large progeny in each generation would gradually cause the species to change, and to change in an adaptive way.[149] The simplicity of the theory* and its extraordinary explanatory power—of which Darwin was already aware—make it the most elegant theory in biology. There was, though, a basic difficulty that arose from the then current ignorance of the nature and causes of the variation between individual members of a species. If, as was generally believed, the characters of parents were simply blended in their children, any new small favourable variation would get diluted away in successive generations; in that situation it was hard to see how Darwin's natural selection could work. We now know that dilution is not a problem because, although the offspring may well show many of the characteristics of both parents, as far as the genes are concerned—and it is the genes that carry almost all the instructions for making the next generation—they are not *blended* but merely *shuffled*. Even a gene whose effects are not apparent in an individual, because they are suppressed by a dominant gene from the other parent, is nevertheless present in the individual's cells and may be passed on to future generations. By a strange irony, while Darwin worried in England, evidence for the shuffling and suppression of genes was being produced by Gregor Mendel in a monastery garden in the Moravian city of Brünn,** though Darwin never became aware of this work.

Although he was the only son in a poor farming family, Mendel was very reluctant to become a farmer.[150] He did so well at school that his parents were persuaded to send him, at the age of 12, to a high school (*Gymnasium*) twenty miles away, but funds were so short that he was entered 'on half-rations' and from time to time his parents would send him food from the farm. When he was 16,

* 'How extremely stupid not to have thought of that!' was Huxley's response.

** Since 1918 known by its Czech name of Brno (pronounced *Burrr*-no). Moravia is a region in the eastern part of what is now the Czech Republic.

a succession of mishaps left his parents unable to pay the fees at all, but he managed to support himself by tutoring. At 18, wishing to become a teacher, he transferred to the Philosophical Institute in Olmütz, supporting himself partly with the help of his younger sister who was willing to sacrifice a share of the family estate which had been put aside for her as a dowry; but funds were still insufficient and on the advice of his physics teacher, who was also a priest, in 1843 at the age of 21 he entered the Augustinian monastery at Brünn as a novice. At 25 he was ordained, and then discovered that he was neither suited to be a parish priest nor able to pass his exams to qualify as a teacher. But, with his Abbot's support, he studied mathematics, physics, chemistry, biology, physiology, and palaeontology, in Vienna, where he acquired much of the knowledge and some of the skills that were to prove so useful later. Returning to Brünn he settled into life as a teacher at the *Realschule*, unqualified but popular with both pupils and colleagues.

Between 1854 and 1863 Mendel did the experiments on peas that became the foundation of the modern science of genetics. During that time he published nothing about them, but early in 1865 he gave two lectures whose contents were reproduced the following year in an article 'On the hybridization of plants' in the *Proceedings of the Brünn Society for the Study of Natural Science*.[151] In the first paragraph of this article, he explained that what led to his experiments was the experience of artificial fertilization, such as that used in ornamental plants to produce new variations in colour, and the striking regularity in the hybrid forms produced by cross-fertilization between plants belonging to different varieties of the same species. That is no doubt true, but his interests must have been wider than his statement suggests, since we know that before he began working on peas he had been trying experiments in his rooms in the monastery, in which he crossed wild mice with albinos to see what colour coats the offspring would have. A visit from the local bishop found such work inappropriate for a monk, and the abbot was obliged to restrict Mendel's research to plants.

By 1854 a great deal of work had been done on hybridization in plants—work which had led to improvements in the plants but had not led to any general laws. As Mendel himself wrote:

> Those who survey the work done in this area will become convinced that among all the numerous experiments made, not one has been carried out to such an extent and in such a way that it is possible to determine the number of different forms under which the offspring of the hybrids appear, or to arrange these forms with certainty according to their separate generations, or definitely to ascertain their statistical relations.

That sentence pretty well defines the daunting task he set himself.

His first problem was to choose a suitable plant. He wanted it to be easy to grow, to be capable of self-pollination and susceptible to artificial cross-pollination, to exist in a variety of forms showing traits that could be recognized without confusion, and to include varieties that would 'breed true'—that is to say, varieties that would, when self-pollinated, produce offspring with identical traits. The common garden pea met all these requirements and also had another advantage—see Figure 56.

In the pea flower, the male organs (the stamens bearing pollen-filled anthers) and the female organ (the pistil with the sticky stigma at its top end and the ovary at its base) are wholly enclosed in the boat-shaped lower petal, known as the keel, until *after* the time that the pollen is ripe and the stigma receptive. If not interfered with, the flower will therefore normally be self-pollinated. To cross-pollinate, Mendel would arrange to have the two varieties he wished to cross in two separate rows. Before the pollen was ripe he would go from plant to plant in the first row, opening

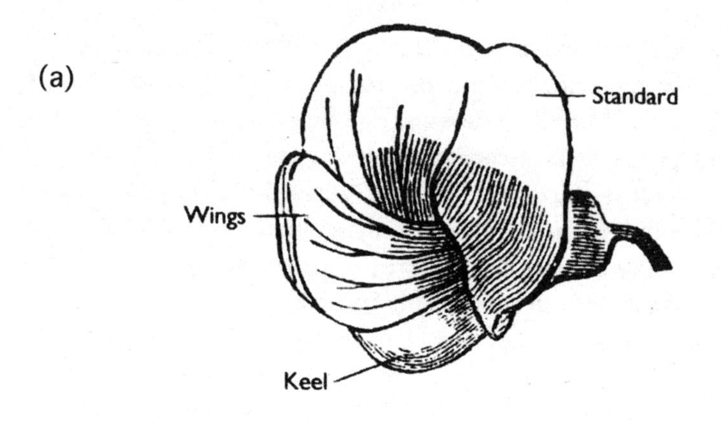

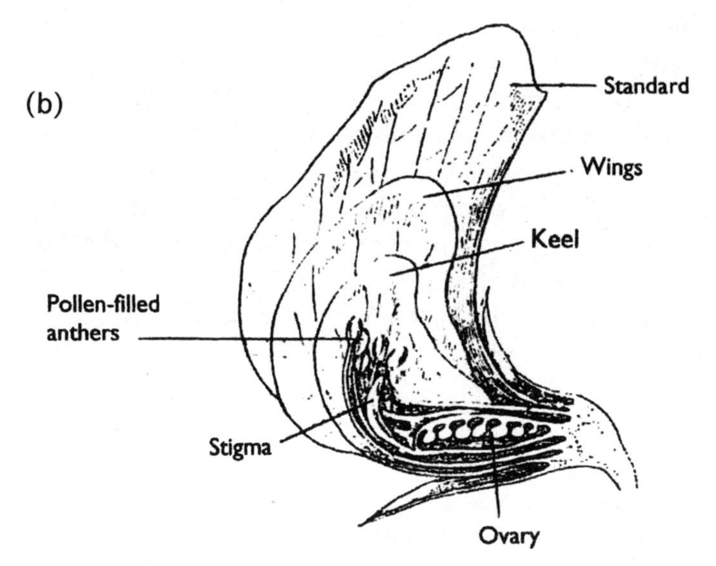

FIG 56 Mendel's pea flowers: (a) the intact pea flower. (b) a vertical section through the flower.

the keel with tweezers, pulling off the anthers, and placing a little calico cap on the flower to prevent access by flying insects. A few days later, when the pollen was ripe, he would attack, one-by-one, the plants in the second row, opening each keel and collecting ripe pollen on a camel-hair brush which he then used to transfer the pollen to the stigma of an emasculated plant in the first row. The calico cap would then be replaced to prevent the pollen being blown off or fresh pollen being brought by a flying insect.[152]

With characteristic meticulousness, Mendel spent two years selecting varieties of pea that bred true, finally choosing twenty-two to use in his experiments. His next task was to choose the character traits whose inheritance he wished to explore. For each trait, his plan was to make hybrids between parents only one of which possessed that trait, and then follow the fate of the trait in successive generations when those hybrids or their descendants were self-pollinated. To simplify decision making, he chose traits that *either* appeared unaltered in the first hybrid *or* did not appear at all—that is, there was nothing halfway. For example, when he crossed a variety that produced round peas with a variety that produced wrinkled peas, he found that all the hybrids in the first generation produced round peas. The trait for wrinkled peas did not appear. He finally settled on seven pairs of traits:

- round peas or wrinkled peas
- yellow peas or green peas
- white flowers or violet flowers
- smooth pods or pods with constrictions between the seeds
- green unripe pods or yellow unripe pods
- flowers distributed along the main stem or flowers bunched at the top
- tall stems (about 2 m high) or dwarf stems (¼ to ½ m high).

In all seven pairs, it is only the first trait that appears in the first hybrid generation. Mendel called such traits *dominant*. Traits that disappear in the first hybrid generation, he called *recessive*—choosing that word because such traits reappear unchanged in the progeny of that generation—as he would show later.

He also showed—though this was not a new discovery—that the nature of the hybrid was the same whether the dominant trait came from the plant supplying the eggs or the plant supplying the pollen. And this was true for all seven of the trait pairs that he investigated.

More interesting results came from observing the consequences of allowing the first generation of hybrids to self-pollinate. For example, when 253 hybrids obtained by crossing 'round pea' and 'wrinkled pea' varieties were allowed to self-pollinate, they produced 7,324 peas, of which 5,474 were round and 1,850 were wrinkled. The ratio of round to wrinkled was very close to 3:1. (Mendel was well aware of the need to use a large number of trials if he wanted reliable statistics.) Similar results were obtained with each of the other trait pairs.

The crucial question then was: why should the ratio be 3:1? Mendel points out that since the character of the hybrid is the same whether the dominant trait comes from the pollen or the egg, it is clear that both the pollen and the eggs must be able to provide what he calls the 'differentiating elements' (*differierenden Elementen* in German)—we now call them genes—that enable the hybrid to show particular traits. And since traits in a parent that do not appear in a hybrid nevertheless do appear in the progeny of that hybrid, *the genes from both parents must still be present in the hybrid.*

Consider the hybrids formed between a true-breeding parent that produces round peas and a true-breeding parent that produces wrinkled peas. Call the gene that enables the plant to make round peas **A**, and the gene that enables the plant to make wrinkled peas **a**. The fertilized eggs from such a cross will contain

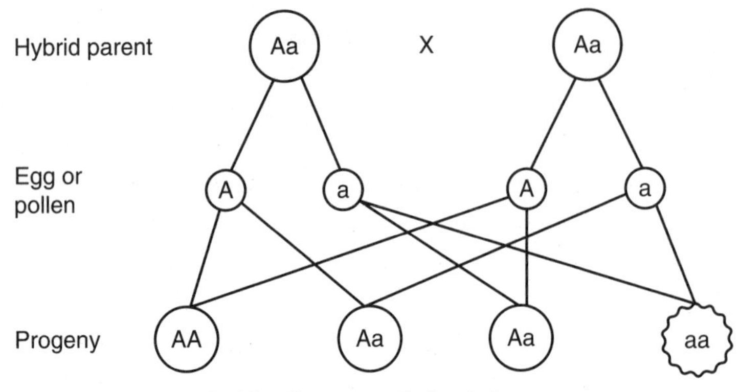

FIG 57 The result of self-pollination of hybrid plants.

both **A** and **a**, but since **A** is dominant and **a** is recessive, the resulting peas will be round—as they were in his experiments.

Now consider what happens when those peas are planted and the flowers on the resulting hybrid plants are self-pollinated. If we assume (what we now know to be true but Mendel didn't) that *an egg or a pollen grain can carry only one kind of gene for any given trait pair,* it follows that each individual egg will contain either **A** or **a**, and so will each individual pollen grain (see Figure 57). If (for both eggs and pollen grains) **A** and **a** are equally likely, it follows that after fertilization there are four equally likely possibilities:

- **AA** where one **A** is from the egg and the other is from the pollen
- **Aa** where **A** is from the egg and **a** is from the pollen
- **Aa** where **A** is from the pollen and **a** is from the egg
- **aa** where one **a** is from the egg and the other is from the pollen.

Because **A** is dominant and **a** recessive, we should expect three round peas for every one wrinkled pea. The fact that the ratio was very close to 3:1 therefore gave strong support to the hypothesis that an individual egg or pollen grain contains only the gene corresponding to one of any given pair of traits. So, in the formation of the egg or pollen grain, there is *segregation* of the genes corresponding to the different members of the pair. This conclusion was later to become famous as *Mendel's law of segregation*.

Equally interesting results came from allowing the first generation of descendants from the hybrids to self-pollinate. When plants from the wrinkled peas (**aa**) that accounted for a quarter of these descendants were allowed to self-pollinate, they all produced only wrinkled peas. When plants from the round peas (**AA** or **Aa**) that accounted for three quarters of the descendants were allowed to self-pollinate, two thirds of them (**Aa**) produced both round and wrinkled peas (in a ratio close to 3:1) and the remainder (**AA**) produced only round peas. Taking the whole of the first generation of descendants from the hybrids together, we can say that a quarter produced only wrinkled peas, half of them produced both round and wrinkled peas in the ratio 3:1, and the remaining quarter produced only round peas. In other words, in the first generation of descendants from the hybrid, a quarter of the plants behaved like one of the original hybrid parents, half behaved like the hybrid itself, and a quarter behaved like the other original parent. And all this could be explained by Mendel's simple model.

In subsequent generations—and Mendel experimented up to the fifth generation of descendants from the original hybrids—a similar pattern held, as his model predicted it would.

This pattern of behaviour not only explained the well known but hitherto puzzling fact that self-pollination of plants of a hybrid strain in successive generations leads to a progressive drop in the proportion of plants showing the hybrid features, and a progressive gain in the proportion showing the features of the parents of the original hybrid; it also enabled Mendel to predict the extent of these changes in each generation. By the tenth generation, only about two in a thousand offspring would resemble the hybrid.

Having worked out the pattern of segregation of parental traits in the progeny of hybrids formed by parents differing only in a single trait, Mendel then asked what happens if the parents differ in two or three traits. He therefore did a number of experiments in which he formed 'dihybrids' or 'trihybrids' and then observed the results of letting them self-pollinate. Because the number of possible combinations was large, these experiments were much more troublesome and predicting their results was more tedious, but they led to a very simple answer. Each pair of traits behaved in precisely the same way as the single pairs of traits in the earlier experiments, and there was no correlation between the behaviour of the different pairs: each went on in its own sweet way. This pattern eventually became known as *Mendel's law of independent assortment of traits* (though, as we shall see later, the behaviour of different traits may be linked if they are controlled by genes on the same chromosome).

As a final *tour de force* Mendel did a number of 'backcrossing' experiments, including the two illustrated in Figure 58. Starting with two true-breeding strains, one with round, yellow peas (i.e. with two dominant traits) and the other with wrinkled, green peas (i.e. with two recessive traits) he made hybrids that necessarily contained genes for both shapes and for both colours. The hybrid's eggs and pollen grains, each of which would contain only one gene for shape and one gene for colour, would therefore be of four kinds: **AB, Ab, aB**, and **ab** (where **A** and **a** determine shape; **B** and **b** determine colour; and capitals are dominant to lower-case letters). He then backcrossed one batch of the hybrids with specimens of the 'dominant parent', and another batch

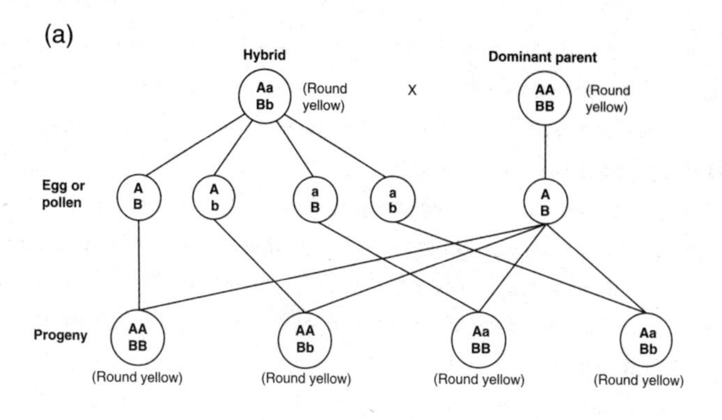

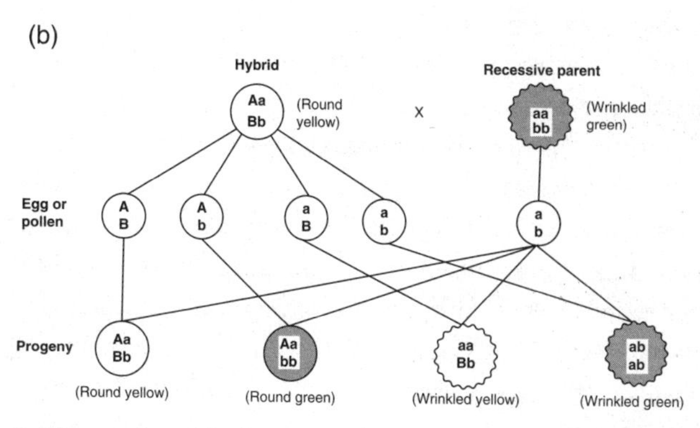

FIG 58 The results of backcrossing hybrids with either parent. *Gregor mendel, the First Geneticist.*

with specimens of the 'recessive parent'. (Meticulous as usual, he did these backcrosses in two ways, using pollen from the parent to fertilize the flowers of the hybrid, or pollen from the hybrid to fertilize the flowers of the parent.) He predicted (i) that the peas produced by backcrossing the hybrid with the 'dominant-parent' would all be round and yellow; (ii) that the peas produced by backcrossing the hybrid with the 'recessive parent' would be of four kinds—round and yellow, round and green, wrinkled and yellow, and wrinkled and green—in roughly equal numbers; and (iii) that it would make no difference whether fertilization was achieved by taking pollen from the hybrid to the parent or from the parent to the hybrid. And all three predictions were right.

Similar experiments with strains differing in flower colour and stem height gave equally impressive results; as did further experiments in which he followed the behaviour of the offspring of these backcrosses when they were allowed to self-pollinate. He also extended his experiments to other species and mostly found similar results, except with hawkweeds, which we now know reproduce asexually.

That such straightforward experiments based on a simple model should provide such understanding of what had long been a recalcitrant subject is a remarkable example of elegance in science. That it was a third of a century before his achievement was recognized is almost equally remarkable. Early in 1867, he sent reprints of his article to perhaps a dozen biologists—though sadly not to Darwin,[†] although he had read and annotated *On the Origin of Species* in a German translation. The absence of any enthusiastic response to his article must have been bitterly disappointing. In March of the following year he was appointed Abbot of Brünn, following the death of the man who had welcomed him as a novice, had encouraged his interests in science, had arranged for him to study in Vienna, had protected him from the strains of parochial duties, and had assisted his botanical studies by building a splendid greenhouse in the monastery garden. Mendel hoped to get back to his plants once he had mastered the tasks expected of

[†] Stories that Darwin had an uncut copy of Mendel's article seem to be apocryphal, as no such copy exists either at Down House or at Cambridge University Library, the two places where his papers are stored. Darwin did have a copy of W.O. Focke's *Die Pflanzen-Mischlinge*, published in 1881, in which the pages referring to Mendel are uncut, and it may be this that gave rise to the story.

him as abbot, but he soon became involved in a protracted and time-consuming dispute about the taxes paid by the monastery to the State, and his research career was at an end. He did, though, repay the generosity of his younger sister by greatly helping her three children.

ACT 2
THE DISCOVERY OF THE ROLE OF CHROMOSOMES

In 1900 three botanists, one in Amsterdam, one in Tübingen, and one in Vienna, independently 'rediscovered' Mendel's 1866 paper, and two years later the English zoologist William Bateson published a monograph that included an English translation of that paper.[153] He also became a strong advocate and defender of Mendel's work, invented the word genetics (from the Greek *genetikos* meaning origin), and showed that sometimes (contrary to Mendel's law of independent assortment of traits) particular traits are inherited together. But major advances in the understanding of heredity had to wait until 1910, when the American Thomas Hunt Morgan had his first success in his studies with the fruit fly *Drosophila melanogaster*.

Born in 1866—the year of Mendel's paper—in Lexington, Kentucky, Morgan came from a wealthy and aristocratic family; his great-grandfather had written the words of the *Star-spangled Banner*, he had an uncle who was a famous Confederacy General, and it was thanks to his father, then the United States consul in Sicily, that the United States became the first nation to recognize Garibaldi's government.[154] Thomas Hunt Morgan, though, was interested in science rather than politics. After graduating from the State College in Kentucky, he went on to do graduate work at Johns Hopkins, and four years later he joined the staff at Bryn Mawr. In 1904 he was appointed to the chair of Experimental

Zoology at Columbia University (New York), and it was here—in the famous Fly Room, that he did the work that led to what became known as 'the chromosome theory of heredity'.

The major weakness in Mendel's work was that he had no idea what sort of thing a 'differentiating element' (gene) was, or how it brought about whatever effects it did bring about. The objects that we now call chromosomes were first observed in the nuclei of dividing plant cells in 1842; later they were also seen in dividing animal cells, and because they are readily stained with basic dyes they were given the name 'chromosome' (from the Greek words *chromos*, meaning colour, and *soma*, meaning body). But in Mendel's day there was no reason to connect them with heredity.

That situation had changed by the time Morgan was starting his experiments with fruit flies. In Leipzig in 1891, Hermann Henking, working on firebugs (*Pyrrhocoris apterus*), noticed an odd-shaped apparently unpaired chromosome in cells that were dividing to form sperm. Similar 'accessory' chromosomes (we now call them Y chromosomes) were later found more widely. Ten years later, Clarence McClung, working with grasshoppers in Kansas, pointed out that because half the sperm had accessory chromosomes, and there are equal numbers of male and female grasshoppers, it was plausible to suggest that the accessory chromosome might be responsible for determining the sex of the offspring.[155] Also early in the new century, Theodor Boveri in Würzburg, and Wayne Sutton at Columbia University, drew attention to the parallelism between the separation of members of chromosome pairs in the formation of eggs and sperm, and the segregation of traits discovered by Mendel.[156] Despite these findings, when Morgan started investigating inheritance at Columbia he still had doubts about Mendelian theory, and also about the role of chromosomes.

Morgan started by working with rats and mice, but in 1908 he switched to fruit flies, which had become popular for breeding experiments. They needed only twelve days to produce the next generation, they could be bred by the thousands in old milk bottles, they lived on rotting fruit, and the cultures needed little maintenance. For two years he bred these flies looking for a new mutation whose inheritance he would then study. To increase the chance of a mutation occurring, he treated the flies in various ways, including exposing them to radiation from radium, but in the end it was an untreated fly that brought him his first success. Fruit flies have red eyes but this fly had white eyes. He then mated this fly, who was a male, to a red-eyed virgin sister who produced offspring all of whom were red-eyed. This seemed analogous to Mendel's crossing of wrinkled and round peas, when the offspring were all round, and it suggested that red eyes were dominant and white eyes recessive. Morgan then mated brother and sister pairs from this hybrid generation and found that in the offspring there were about three times as many with red eyes as with white eyes. This was, of course, analogous to Mendel's finding of three round peas to one wrinkled pea, when he allowed the first generation of hybrids to self-pollinate, and it strongly supported the notion that red eyes were dominant and white eyes recessive. So far, so good.

There was, though, another and startling finding. In the generation showing the 3:1 ratio of red eyes to white eyes, all of the white-eyed flies were male; of the red-eyed flies, about two thirds were female and one third male. Later Morgan and his colleagues discovered two further spontaneous mutations—one causing rudimentary wings, and the other a yellow body—and these too were linked to the sex of the fly. A possible explanation was that the genes responsible for eye colour, rudimentary wings, and yellow body are all carried on the same chromosome, and that this chromosome is also involved in sex determination.

The obvious next step was to look at the fruit fly's chromosomes under the microscope. There are four pairs of chromosomes in each cell nucleus, one member of each pair coming from the male parent and the other from the female parent. But what Morgan found was that only three of the pairs consisted of similar looking chromosomes; the remaining pair consisted of two normal looking chromosomes in the female, but in the male one chromosome looked normal, while the other was smaller and odd. Using symbols to label the chromosomes of the fourth pair, that pair can be written XX in the female and XY in the male. When the members of chromosome pairs separate during the formation of eggs or sperm, all eggs will therefore carry one X chromosome, and so will 50% of the sperm; the remaining sperm will carry a Y chromosome. When eggs are fertilized by sperm, it follows that about half the fertilized eggs will be XX and will produce female offspring, and the remainder will be XY and will produce male offspring. This provides a neat explanation of the equal numbers of male and female offspring.

But what about the unexpected linkage between sex and eye colour? Looking at the situation from the point of view of the offspring, the female offspring receives one X chromosome from each parent; the male offspring receives an X chromosome from his mother and a Y chromosome from his father. Morgan immediately saw that *if the X chromosome carries the (recessive) gene for eye colour but the smaller Y chromosome does not*, this would explain the strange correlations between sex and eye colour that he had observed. A male that had the gene for white eyes in his X chromosome would have white eyes. A female would have white eyes only if she inherited the gene for white eyes from both parents. (This is, incidentally, just like the inheritance pattern of the commonest form of colour blindness in people.)

Morgan's hypothesis not only provided an elegant explanation of the unexpected observations; it also, and for the first time, linked the cause of a particular trait to a particular chromosome. What Mendel would have called a differentiating element was at last anchored to an object.

And Morgan went further. If genes for several different traits were all on the same chromosome, you would expect those genes to be inherited together, and usually they were. But occasionally, he noticed, some genes for *linked traits* would separate, though other genes on the same chromosome would not. He explained this by referring to a then controversial paper published by Frans Janssens, a professor in Leuven, in 1909.[157] Janssens was an expert microscopist who had watched the development of germ cells (eggs or sperm) in amphibians and described, correctly as it turned out, a very surprising phenomenon. At an early stage in the development of these cells, when members of each pair of chromosomes come together they tend to twine round each other and may break and rejoin so that equal and corresponding parts of the two chromosomes may be exchanged—a process that became known as 'crossing over'—see Figure 59. A result of this exchange is that the linkage between traits caused by genes originally on the same chromosome may be broken.

Morgan realized that the chance of two genes on the same chromosome getting separated as a result of crossing over would be small for genes whose positions on the chromosome were close together and would increase as the distance between them increased. One evening in 1911, Alfred Sturtevant, a 19-year-old Columbia student who was helping Morgan, used his knowledge of variations in the strength of the linkage between traits caused by genes on the same chromosome to estimate the relative distances between those genes. Neglecting his homework, he sat up half the night and produced the world's

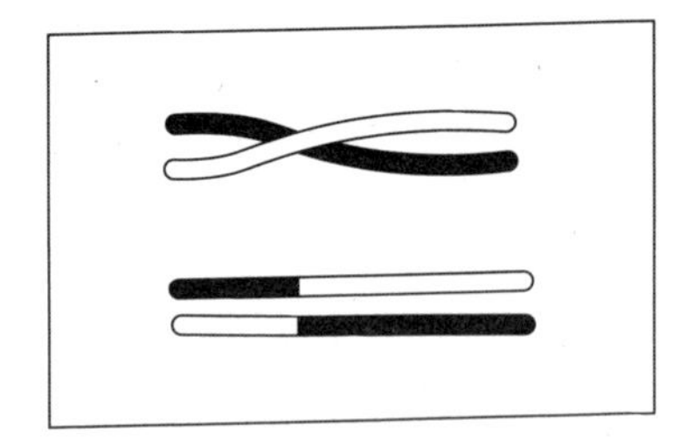

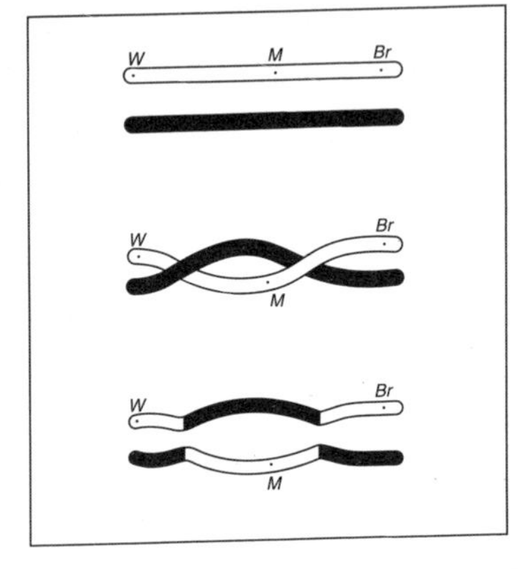

FIG 59 'Crossing over' in chromosomes.

first chromosome map.[158] It had only three genes on it, but it was the first step in the long series that led to the mapping of the human genome.

By 1914, more than fifty *Drosophila* mutants had been discovered, and Morgan and his colleagues showed that they fell into four clear linkage groups, corresponding to the four chromosome pairs that were found in all *Drosophila melanogaster* cells. By the early 1920s, Sturtevant and another of Morgan's former research students had shown that two other species of *Drosophila*, which had respectively five and six pairs of chromosomes, also had respectively five and six linkage groups.[159] This parallelism between the number of chromosome pairs and the number of linkage groups provided a most elegant proof of the chromosomal theory of heredity.

ACT 3
WHAT SORT OF MOLECULES CARRY INFORMATION FROM GENERATION TO GENERATION?

In other words: what are genes made of? From the 'crossing-over' experiments, it was clear that genes were substances of some kind that exist in linear arrays in chromosomes. By the 1930s, it was known that chromosomes consist largely of a mixture of nucleic acids and proteins; and because proteins were known to be extraordinarily varied in structure, being built up with twenty different kinds of amino acid, while nucleic acids were thought to have a rather monotonous structure, it was generally assumed that it was the proteins that were largely responsible for determining what individual genes do.

This view was changed dramatically by a paper published by Oswald Avery and his colleagues in 1944 that was entirely concerned with experiments with pneumococci[160]—the bacteria that cause pneumonia in both people and animals. Avery, then aged 67, had been born in Canada to an English Baptist clergyman, but the family moved to New York when Oswald was ten, and he remained there for the whole of his working life, first qualifying as a physician and then switching to bacteriology and immunology.[161]

What led to Avery's famous 1944 paper was a startling observation made sixteen years earlier by Frederick Griffith, a medical officer in the Ministry of Health in London.[162] It was already known, largely from Avery's earlier work, that pneumococci come in several immunologically distinguishable types (Type I, Type II, etc.), and that each type can exist in two strains: **S** strains,

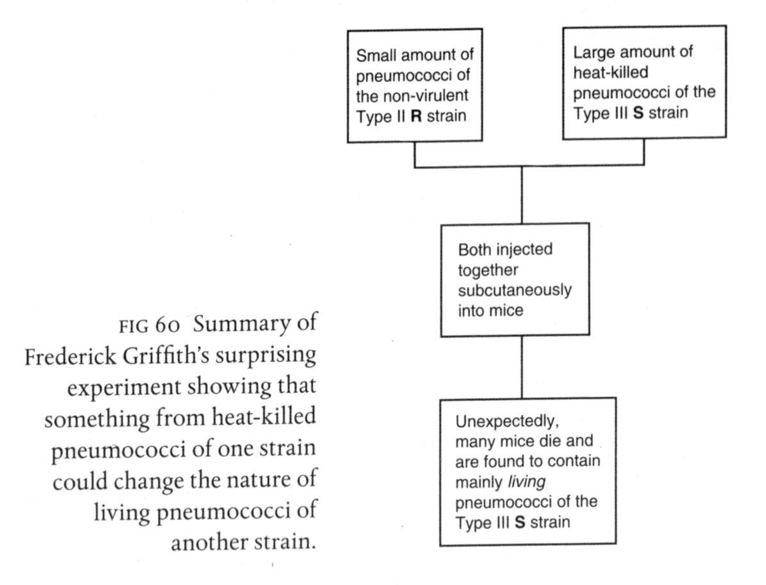

FIG 60 Summary of Frederick Griffith's surprising experiment showing that something from heat-killed pneumococci of one strain could change the nature of living pneumococci of another strain.

so-called because they produce **smooth** shiny colonies when cultured, and **R** strains, which produce **rough** dull-looking colonies. The individual bacteria in the **S** strains are each enclosed in a capsule built from sugar molecules or derivatives of sugar molecules, and they are virulent because the capsule protects the bacterium from being eaten by the host's immune cells. The **R** strains have no capsules and are not virulent. What Griffith did was to inject mice, through their skin, with a *small* amount of a *living* **R** strain derived from pneumococcus Type II, together with a *large* amount of a *heat-killed* **S** strain derived from pneumococcus Type III—see Figure 60.

Since neither the living **R** strain nor the heat-killed **S** strain were lethal to mice, he did not expect the mice to die; but many of them did. What is more, blood taken from the hearts of the dead mice contained mainly *living* Type III pneumococci of the **S** strain—pneumococci which, when put on a suitable culture medium, would grow generation after generation of pneumococci of the same type and strain. Since there was no doubt that the material injected into the mice had contained no *living* bacteria of this kind, it seemed that, unless one believed in resurrection of the dead, the heat-killed pneumococci must have produced something that transformed the living pneumococci of the Type II (**R** strain) into the living Type III (**S** strain).

Avery, like many working in the field, was sceptical about Griffith's claim, but it was soon confirmed by others. And it turned out that it wasn't necessary to use mice as a host; the transformation could be demonstrated simply by mixing the heat-killed **S** strain and the living **R** strain on a glass dish. By 1932, James Alloway, a colleague of Avery's at the Rockefeller Institute, showed that you didn't even need *intact* pneumococci of the heat-killed Type III **S** strain.[163] He broke open the capsules using bile salts—detergents were not widely used until after the Second World War—extracted the contents with a salt solution, and got rid of the solids, including any unbroken cells, by passing the whole suspension through a fine filter. The resulting sterile solution was sufficient to cause the transformation of the living R strain. At last, Avery was convinced. The solution must contain what he called the 'transforming principle'—shades of Mendel's 'differentiating element'.

Being convinced, he was determined to discover what the 'transforming principle' was, but there was a personal difficulty. In the early 1930s, Avery was in his mid-50s and suffering from Graves' disease—an unpleasant condition caused by overactivity of the thyroid gland—and for several years his role in research was limited to observing and advising his junior colleagues. By the late 1930s, though, he was fully active again, and in 1940 he and his colleagues began to tackle what was a formidable task; formidable, because the solutions containing the transforming principle also contained an uncertain, though undoubtedly large, number of other constituents. The only way to identify the active one was, first, to exclude some by showing that their removal or destruction did not affect the potency of the solution, and then to decide between the remaining candidates by looking for an incriminating relationship between the relative concentration of each substance in a series of solutions and the relative efficacy of those solutions in producing transformation. A further complication was that pneumococci contain an enzyme capable of inactivating the transforming principle.

In the next four years, Avery and two colleagues succeeded in solving the problem. Their method of purifying the transforming principle was based on that used by Alloway, and they solved the problem of the indigenous inactivating enzyme by starting with ice-cold pneumococci, warming them very quickly to 65° and then keeping them for half an hour at that temperature. The rapid rise in temperature ensured that the troublesome enzyme was destroyed before it had time to do much harm. The subsequent procedures were too tedious to be described here, but Avery and his colleagues ended up with a solution that contained the elements nitrogen and phosphorus in a ratio that fitted well with the composition of nucleic acids. Tests for protein were negative, and enzymes that break down protein had no effect on the transforming activity of the solution. In contrast, a chemical test for deoxyribonucleic acids (DNA) was positive, and an enzyme that breaks down DNA abolished transforming activity. An enzyme that breaks down ribonucleic acids (RNA) was ineffective.

A number of physicochemical studies—absorption of ultraviolet light, rate of sedimentation in an ultracentrifuge, rate of movement in an electric field—also showed striking parallels between the known features of DNA and the observed features of the material in the largely purified solution.

Finally, Avery and his colleagues measured the transforming activity of successive dilutions of the purified solutions, and showed that transformation of the pneumococci could be produced by concentrations (expressed as dry weight of DNA per weight of solution) as low as 1 part in 600 million.

Potent stuff indeed! But what makes this paper by Avery and his colleagues so stunning is not the potency of DNA but what they proved DNA can do. It is not merely an ingredient in the transformation, nor is it merely a catalyst. To enable pneumococci of the Type II (**R** strain) to make pneumococci of the Type III (**S** strain), the DNA must provide the information to make all the bacterial proteins characteristic of the Type III (**S**) strain; and it must continue to do this job in generation after generation—so it must also provide the information necessary for its own reproduction. As Avery wrote to his bacteriologist brother, 'Sounds like a virus—may be a gene.'[164] And he was, of course, right. Two years later, when Avery was given the Copley medal of the Royal Society, the President, Sir Henry Dale, commented:

> Here surely is a change to which, if we were dealing with higher organisms, we should accord the status of a genetic variation, and the substance inducing it—the gene, one is tempted to call it—appears to be a nucleic acid of the desoxyribose type.[165]

Dale's hesitation about using the word 'gene' reflects the then current belief among some biochemists and biologists that bacteria did not possess genes.[166]

You may of course say that this is all very impressive, but was their work elegant? After all, the surprising finding was that of Griffith sixteen years earlier; the essentials of the purification were worked out by Alloway; and what Avery and his colleagues did in their 1944 paper was to apply a whole lot of standard and tedious techniques until they identified the culprit. But this would be unfair to Avery. Firstly, his whole approach depended on a succession of his earlier elegant papers, in which he showed how differences in the appearance, virulence, and responses to the hosts' defence mechanisms of various known strains of pneumococci depended on their chemical composition. Secondly, to prove the crucial (and at that time unexpected) role of DNA in heredity through a succession of simple experiments, examining just one kind of behaviour shown by just one kind of bacterium, is breathtakingly economical. The proof is simple, ingenious, concise, and persuasive. It is also very satisfying. What Avery and his colleagues did in the 1940s may have been a few small steps for a bacteriologist, but it was certainly a giant leap for mankind. And in the long run—indeed even in quite a short run—discovering the basis of heredity is much more significant than putting a man on the moon.

ACT 4
HOW IS INFORMATION ENCODED IN THE DNA, AND HOW IS IT USED?

If DNA had the role that the experiments of Avery and his colleagues revealed, it had to be capable of supplying a vast amount of information, and it clearly couldn't have the simple uniform structure it was then usually assumed to have. It was known to consist of units each possessing a phosphorylated five-carbon sugar molecule (deoxyribose) attached to one of four possible nitrogenous bases—adenine, guanine, thymine, and cytosine. It had originally been supposed (for no very good reason) that the molecule contained only four of these units, one with each kind of base, and though that notion had to be given up when it became clear that the molecular weight of DNA was very large and the molecules were very long, it was not at all clear what the structure was.

Fascinated by Avery's work, Erwin Chargaff, a biochemist in his late 30s at Columbia University, decided to compare the composition of DNA from various different species, breaking the DNA into its various components and making use of the recently developed technique of paper chromatography to separate those components.[167] He found, as he expected, that the ratios of the four different bases differed markedly between DNA from bacteria, yeast, ox, pig, sheep, and humans. But he also found—what was quite unexpected—that in any one organism, adenine and thymine occurred in very similar proportions, as did guanine and cytosine. In the following year John Griffith (nephew of the Frederick Griffith who had discovered transformation in pneumococci) produced quantum-mechanical arguments suggesting that there might well be selective attraction between adenine and thymine and between guanine and cytosine, so the idea of 'base pairing' was around.[168] And both Griffith and Crick realized that such pairing might indicate that replication was complementary: in other words, that a sequence of bases would lead to the formation of a sequence of the preferred partners of those bases. But at that time pairing was thought of as the stacking of the paired bases one above the other—possibly interleaving bases from different strands of the DNA molecule—and it was not until 1953 that Watson and Crick produced their model, with pairs of bases (linked by hydrogen bonds) in the same plane, but in different strands.

That model was created in 1953, and was largely based on the X-ray diffraction studies of DNA by Rosalind Franklin and Maurice Wilkins and their colleagues at King's College, in London. Maurice Wilkins, who initiated this line of research at King's, says in his autobiography that he had been stimulated by early knowledge of Avery's work, and it was by meeting Wilkins at a conference in Naples in 1951 that Jim Watson became interested in DNA.

In the chapter on wave theory,[169] earlier in this book, we saw how in his famous 'two-slit experiment' Thomas Young used the distance between the bands in the diffraction pattern to calculate the wavelength of the light. That calculation depended on knowing also the distance between the two slits, and the distance between the slits and the screen on which the pattern was seen.[‡] Had Young known the wavelength of the light and the distance of the screen from the slits, he would have been able (by doing the calculation in reverse) to calculate the distance between the slits. More generally, light of known wavelength diffracted from any regular repeating structure of appropriate size, and focused on a photographic film, causes a pattern of lines or dots whose distribution makes it possible to calculate the dimensions of the structure—the equivalent of 'the distances between the slits'.

In 1812, Max von Laue,[170] a German theoretical physicist working in Munich, was talking to a colleague in the English Garden—the garden created by Rumford. The conversation was about the colleague's work using visible light to study the structure of crystals. At that time it was suspected that crystals had a regular atomic structure, and also suspected that X-rays might be electromagnetic rays, like light rays but with shorter wavelengths. It occurred to Laue that if both suspicions were right it ought to be possible to produce diffraction patterns by exposing crystals to a beam of X-rays. He discussed this idea with his colleague, and with other faculty members, and two weeks of experimenting with crystals of copper sulphate proved he was right. This not

[‡] This is because the basis of the calculation is the assumption that if the difference in the distances between a given point on the screen and the two slits is an odd number of half-wavelengths the light reaching the point from the two slits will be exactly out of phase and will cancel out; if it is an even number of half-wavelengths the light from the two slits will be exactly in phase and the point will be brightly lit.

only showed that both suspicions were true, but also led to the new field of X-ray structural analysis, with William and Lawrence Bragg, father and son, leading the way. Einstein described the experiment as one of the most beautiful in physics, and Laue was awarded a Nobel Prize in 1914.

Gradually, investigators became more ambitious and began looking at biological materials. In 1928 William Astbury, in Leeds, was studying the X-ray diffraction of keratin, the protein in wool; in 1934 Desmond Bernal showed that crystals of the enzyme pepsin would yield clear-cut X-ray diffraction patterns; and by 1937 Max Perutz was beginning to work with crystals of haemoglobin. The trouble with very large complicated molecules, though, was that there was no straightforward way of determining the structure from the diffraction pattern. Certain patterns might suggest particular regularities and particular dimensions in the crystal, but to progress further all you could do was to use whatever information you had about the substance being investigated to design hypothetical structures, and then see whether the observed diffraction pattern fitted the pattern expected from those structures.

Interesting but, as it later turned out, incorrect conclusions about the structure of DNA had been obtained by Astbury, in Leeds in 1938, and by Sven Furberg, who joined Bernal's department at Birkbeck College in 1947. In May 1950, though, the Swiss biochemist Rudolph Signer gave Maurice Wilkins a sample of calf thymus DNA that had been prepared using very gentle methods. Using this sample, which tended to form very long thin fibres that looked uniform and transparent under the microscope, Wilkins found that he could get much sharper X-ray diffraction patterns than those obtained by his predecessors. In 1951, Rosalind Franklin joined the group at King's and showed that DNA diffraction patterns were confused because the molecule could exist in two forms depending on the humidity; by controlling the humidity, she was able to avoid the confusion and investigate both forms. The gradual development of knowledge of the structure of DNA between 1951 and the publication of the Watson–Crick model and accompanying papers by workers at King's, in April 1953,[171] is discussed at length in Olby's *The Path to the Double Helix*,[172]and in Judson's *The Eighth Day of Creation*.[173] A briefer account, but prepared with the advantage of access to Rosalind Franklin's lab notebooks, is given in an article written by Aaron Klug in 2003 to celebrate the fiftieth anniversary of the birth of the model.[174] The story is surprisingly complicated, often highly technical, sometimes highly personal, and usually fascinating, but because this is a book primarily about elegance rather than history I shall not attempt to summarize it. Instead, I want to discuss the way in which the Watson–Crick model not only provides a thoroughly convincing account of the structure of DNA, but also—and quite unexpectedly—an equally convincing account of the way it does its job. This must be the ultimate example of 'buy one, get one free'.

But first the structure. What Watson and Crick proposed was that DNA was a double helix consisting of two intertwined sugar-phosphate chains with the nitrogenous bases projecting inwards from each of the chains. The molecule can be thought of as resembling a twisted ladder, in which each upright consists of a chain of identical sugar molecules linked by phosphate groups, and the rungs consist of paired nitrogenous bases—either adenine and thymine, or guanine and cytosine. A portion of the structure is drawn uncoiled in Figure 61 to show the sugar-phosphate chains running along the outside of the molecule and the paired nitrogenous bases forming the rungs. The figure is misleading in one way, though, since it looks as if the bases and the sugars all lie in the same plane, whereas in fact the planes of the bases lie at right

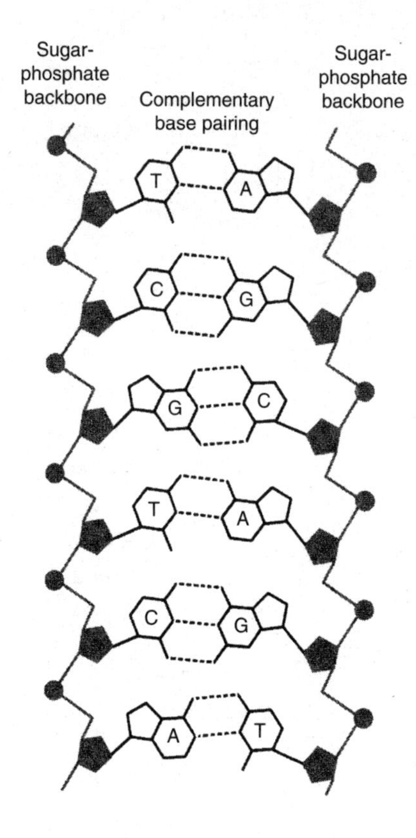

FIG 61 The structure of DNA. A portion of the double helix is drawn uncoiled.

angles to the axis of the helix, so that the base pairs are stacked up along this axis and about 0.34 nanometres apart. The diameter of the helix is about 2 nanometres and each coil of the helix contains ten base pairs.

Each sugar is linked to the adjacent phosphate groups and to the adjacent nitrogenous base by ordinary strong covalent bonds, but what holds together the two bases in each pair are much weaker *hydrogen bonds*. (They are represented by the dashed lines connecting the two halves of the molecule in Figure 62.) These weak bonds are the result of the weak electrical attraction that exists between

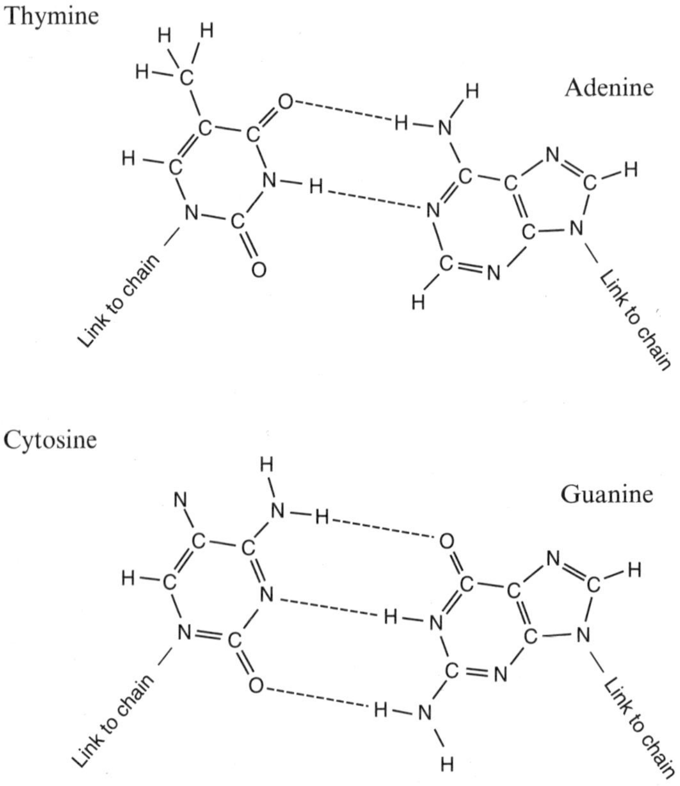

FIG 62 The pairing of bases by hydrogen bonds.

the very small positive charge possessed by hydrogen atoms attached to a nitrogen atom in one member of the pair, and the very small negative charge possessed by oxygen or nitrogen atoms in the other member of the pair—see Figure 62.

A role for hydrogen bonding had been suspected before, but Watson and Crick recognized two further subtleties. *Firstly*, the geometry of the bases was such that two hydrogen bonds would link together adenine and thymine, and three would link guanine and cytosine; and *secondly*, the geometry of the whole structure was such that the positions of the points at which the base pairs in a rung of the ladder are attached to the sugars in the uprights of the ladder would be almost identical whether the rung (going from left to right) was adenine–thymine or thymine–adenine, or guanine–cytosine or cytosine–guanine. This suggested that, so far as geometrical constraints are concerned, all four possibilities would be equally likely, so that any sequence of pairs along the double helix would be possible.

The ability of the Watson–Crick model to account for the detailed X-ray diffraction patterns of DNA, and for the unexpected Chargaff ratios, made it attractive enough; but there was a further feature that made it even more attractive. Because of the restricted base pairing, the sequence of bases attached to one helical sugar-phosphate chain (one upright of the twisted ladder) was complementary to the sequence of bases attached to the other helical sugar-phosphate chain (the other upright of the twisted ladder). This implied that if you pulled the two halves of the double helix apart—and we have seen that the bases in each pair are held together only by weak hydrogen bonds—you would have two separate strands *each of which would contain all the information necessary for it to act as a template for building a copy of the other*—see Figure 63. Such replication is sometimes called *semi-conservative* because each parent strand continues to exist along with a daughter strand.

Of course, the hypothesis that this was the way inheritance worked left a succession of unanswered problems. What sort of machinery separated the strands or built a new strand using an original strand as the template? What sort of code allowed the sequence of bases in a strand of DNA to be translated into a sequence of amino acids in a polypeptide chain in a protein? Given the existence of such a code, what sort of machinery selected

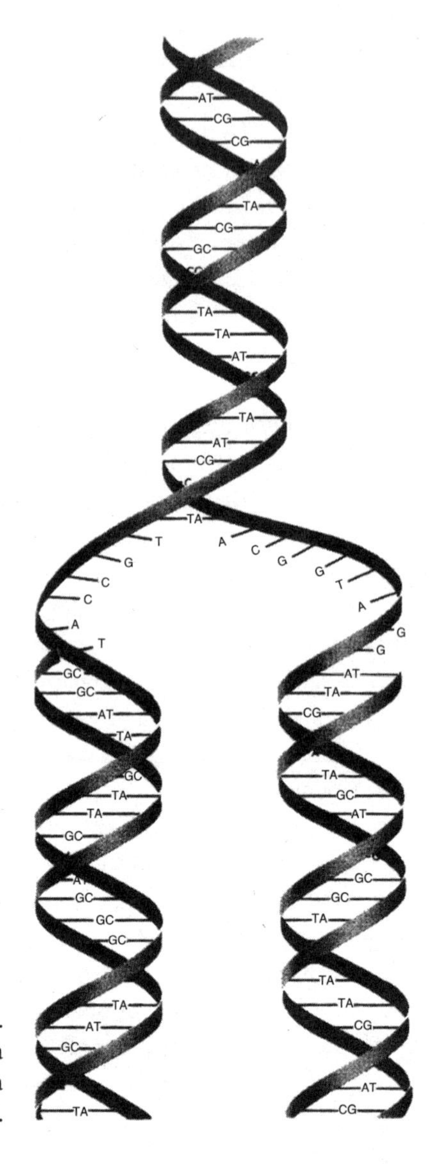

FIG 63 Replication of DNA. Each strand is used as a template to form a complementary strand.

the amino acids and joined them together to form a polypeptide chain? And how is the whole process controlled? After all, the DNA is the same in all cells in a particular organism, but your pancreatic cells don't make haemoglobin and your developing red blood cells (which still contain DNA) don't make insulin. In 1953, though, all these question were for the future. The current feeling, among those who were aware of the situation, was that, for a molecule whose whole purpose is to reproduce itself and to convey a great deal of information from itself to its offspring, the structure proposed by Watson and Crick provided an explanation that was simple, ingenious, concise, and persuasive, with an unexpected aspect to it, and very satisfying. It was, in other words so elegant that it seemed bound to be true. And of course it turned out that it was true—though, as we shall see later, to argue from elegance to truth can be frighteningly unsound.

Before we look at the unreliability of elegance, though, I want to describe one last elegant experiment—the experiment that proved that the replication of DNA really was semi-conservative, i.e., that each parent strand continued to exist along with the daughter strand formed from it. The experiment was done in 1958 by Matthew Meselson and Franklin Stahl, at the California Institute of Technology, using the bacterium *Escherichia coli*.[175] To detect whether individual strands of DNA were conserved, they needed to be able to label parental strands so that they could be distinguished from newly formed daughter strands. They did this by growing the bacterium for fourteen generations in a medium in which the only source of nitrogen was ammonium chloride in which the nitrogen atoms had an atomic weight of 15 (^{15}N) instead of the usual 14 (^{14}N). After so many generations, virtually all the N atoms in the DNA would be ^{15}N and the density of the DNA would be about 1% higher than normal. They then transferred the bacteria abruptly to a similar medium containing normal ammonium chloride so that if Watson and Crick were right *a daughter strand of DNA formed after the transfer* would contain only ^{14}N. The crucial question was: how does the distribution of ^{15}N and ^{14}N between different molecules of DNA change after successive rounds of replication?

To answer this question, they took advantage of the increased density of DNA containing only ^{15}N in its DNA. They centrifuged extracts of the bacterial DNA in a solution whose density was not uniform but graded—being slightly less than the density of DNA at the top of the centrifuge tube and slightly more than the density of DNA towards the bottom of the tube. In this situation, the DNA would move until its density matched that of the surrounding fluid, when it would stop. The final position of the DNA would, therefore, indicate its density, and that final position could be determined by taking an ultraviolet absorption photograph of the tube; the DNA showed up as a dark band.

What they found was that normal DNA containing ^{14}N and DNA containing only ^{15}N gave distinct narrow bands, the distance between them corresponding, as expected, to a density difference of about 1%. The interesting question was how did the density of DNA change in successive generations when bacteria that had started with 100% ^{15}N were suddenly transferred to a medium containing only ^{14}N?

The answer was that after one generation the DNA gave a narrow band halfway between the ^{15}N band and the ^{14}N band. This is exactly what the Watson–Crick model predicts, since each double helix will contain one parental strand and one daughter strand—see Figure 64.

After two generations, the DNA showed two bands of about equal strength, one similar to that found after one generation, and the other similar to the ^{14}N band. Again this is just what the model predicts. After three generations there were the same two

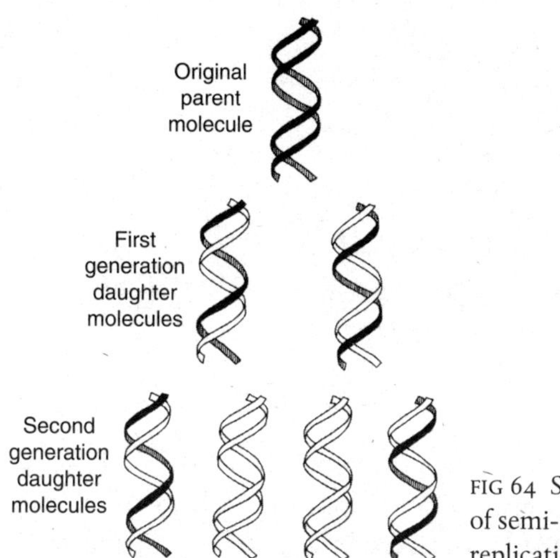

FIG 64 Schematic diagram of semi-conservative replication.

bands but the ^{14}N band was much stronger than the 'halfway' band—which makes sense because the double helixes containing one parental strand will form a smaller proportion of the population. After four generations, almost all of the DNA was in the ^{14}N band. A crucial negative finding is, of course, that at no time were the DNA bands at any other positions.

This Meselson–Stahl experiment has been called 'the most beautiful experiment in biology'.[176] If there has to be a beauty contest, I think I should back Alan Hodgkin's proof of the local circuit theory of nerve conduction (see page 163), though the Meselson–Stahl experiment is certainly very elegant, and might have been a suitable ending to this book.

But I want to finish with a cautionary tale.

10

EPILOGUE: A CAUTIONARY TALE

As we have seen, genetic information is held in DNA as long sequences of nitrogenous bases, which are of only four kinds. Almost everything going on in the body depends on proteins, and proteins are made up of long chains of amino acids, which are of twenty different kinds. A crucial problem in the 1950s, then, was how could sequences of four kinds of unit define sequences of twenty kinds of unit? If each of the four bases specified a single amino acid, only four amino acids could be specified. If two consecutive bases specified each amino acid, it would be possible to specify $4^2 = 16$ amino acids. That would still be too few. To specify twenty different amino acids purely from the sequence of bases, it would be necessary to use three bases to specify each one; that would allow $4^3 = 64$ different amino acids to be specified. But if the coding system allows sixty-four different amino acids to be specified, isn't it odd that we have only twenty?

And there was another problem. If each amino acid is specified by a triplet of bases—a codon as geneticists now call it—the protein-synthesizing machinery has to know where each codon starts and stops. The strand of nucleic acid that is acting as a template contains a very long row of bases, but they are not

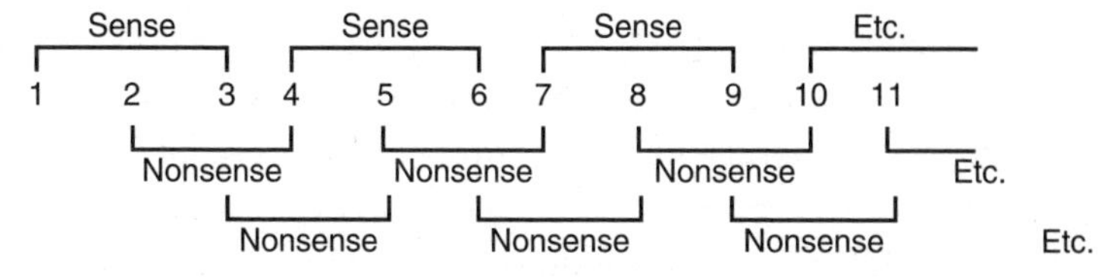

FIG 65 The requirements of the comma-less code.

grouped in any way, nor is there any obvious punctuation. No doubt the machinery starts at a particular place on the nucleic acid strand, and finishes at another particular place, but in between does it just advance three bases at a time, adding on the appropriate amino acid at each step? Or do certain groups of bases serve as punctuation? In 1957, Francis Crick, John Griffith, and Leslie Orgel produced an extremely elegant hypothesis that seemed to solve both these problems.[177] It was a scheme that looked infallible yet did not require punctuation; and they called it the comma-less code or comma-free code. The idea was ingenious and absurdly simple.

With sixty-four possible triplets, we need only a minority of them to specify amino acids. The rest could be meaningless—'without sense' as people who think about codes would put it. Is it possible, then, to choose a group of meaningful triplets such that, when they are placed end-to-end *in any order* no overlap between them yields a group of three bases that has a meaning? The diagram in Figure 65 shows what we want. The numbers on the top line represent successive bases in a nucleic acid strand. Consider the triplets in the top line: 123, 456, 789, and so on. We want all these triplets to designate individual amino acids, but we also want all the triplets indicated in the lower lines to be meaningless. If that could be achieved there would be no danger in getting slightly out of step, and putting in the wrong amino acid in that step and in subsequent steps; any misread triplet would be meaningless.

It turns out that it is theoretically possible to achieve the arrangement represented in the diagram, provided that the number of meaningful triplets needed is not too great; and if you ask how many you can have, the answer is twenty. As Crick put it, writing much later, 'You fed in the magic numbers four (the four bases) and three (the triplet) and out came the magic number twenty, the number of different amino acids.'[178]

If you look at the December number of the *Scientific American* for 1959, you will find a long article by Mahlon Hoagland about protein synthesis.[179] In it he considers 'how the four component units of DNA (and RNA) can determine the order of twenty units in protein molecules' and he points out that 'one ingenious solution' to this problem is Crick, Griffith, and Orgel's suggestion of the comma-less code. He does not mention any doubts about their suggestion, and I suspect that most readers of the article in 1959 would have come away with the impression that the comma-less code was probably sound. The originators of the code, who in response to general interest had published their hypothesis in 1957, were more cautious, which was wise of them, for it turned out to be entirely wrong. Far more than twenty of the sixty-four possible triplets code for amino acids, with the result that many amino acids are coded for by more than one triplet. It seems that, without any punctuation other than an initial *start* and a final *stop* sign, the protein-synthesizing machinery is capable of moving reliably along the nucleic acid strand recognizing each successive triplet of bases as an instruction to add a particular amino acid to the growing polypeptide train.

The moral of this story is that it's fine to get pleasure from elegant theories and elegant experiments, and it's fine to strive to create such theories and do such experiments—but don't get seduced by elegance: an elegant theory is not necessarily true. As the philosopher Peter Lipton put it: 'The 'loveliest' explanation is not necessarily the 'likeliest'.'[180]

APPENDIX TO CHAPTER 4

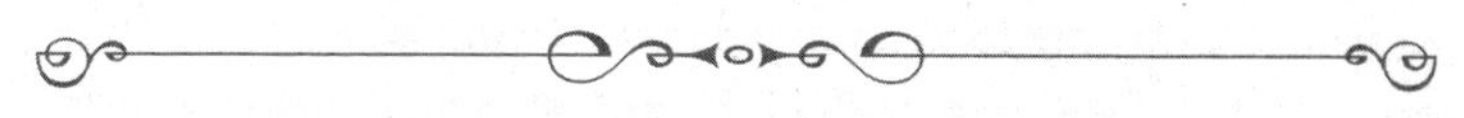

THE SECOND LAW OF THERMODYNAMICS

Heat won't pass from a cooler to a hotter.
You can try it if you like, but you'd far better notta
Flanders and Swann, 1963

In the tailpiece to the chapter on heat we saw that although the law of conservation of energy (also known as the first law of thermodynamics) places no limits on the transformation of energy in any one form into energy in any other form, there is a curious asymmetry. The *complete* (or almost complete) transformation of mechanical energy or electrical energy or chemical energy into heat is easily achieved. The *complete* transformation of heat into mechanical or electrical or chemical energy is not possible. Oddly, the first and major step in understanding this asymmetry—a step that was to lead to the discovery of the second law of thermodynamics—was taken about twenty years before the first law was discovered. It was taken by a young Frenchman, Sadi Carnot, in a remarkable thought experiment.[181]

Carnot was born in 1796 in the Luxembourg Palace, in Paris, where his father, a military engineer and strategist who had been enormously successful during the revolutionary war, was a member of the five-man Directory that then governed France. There is a story that the 4-year-old Sadi Carnot was being looked after by Madame Bonaparte and other ladies in a rowing boat, when Napoleon appeared and 'splashed water on the rowers by throwing stones near the boat. Sadi... watched for a while, then indignantly confronted Bonaparte, called him "beast of a First Consul" and demanded that he desist. Bonaparte stared in astonishment at his tiny attacker, and then roared with laughter'.[182]

Like his father, Carnot was concerned with working out the basic principles in the design of machinery, and his particular interest was steam engines. By the time he was in his 20s, such engines had existed for over a century, and during that period there had been very considerable improvements in efficiency. The first practical engine, invented by Thomas Newcomen in 1712, had a vertical cylinder and a piston that was counterweighted so that little force was needed to raise it. Steam at quite low pressure was admitted to the bottom of the cylinder, and raised the piston to the top. Cold water was then sprayed into the cylinder to condense the steam, greatly reducing the pressure in the cylinder so that atmospheric pressure pushed the piston down to the bottom, ready for the next cycle. The trouble with this system was that in each cycle the temperature of the cylinder went from low to high and back to low, so a large fraction of the heat being put into the system was being used up heating the metal of the cylinder. In 1769 James Watt solved this problem by adding a separate smaller chamber in which the steam was condensed, so that the main chamber, which was insulated, could be kept hot all the time. He also used higher-pressure steam, a horizontal cylinder, and an arrangement that allowed the steam to be admitted first on one side of the piston and then on the other.

In 1824 Carnot wrote a 118-page essay entitled *Reflections on the Motive Power of Fire and on the Appropriate Machines for Developing that Power.*[183] In it he attempted to answer these questions:

- How much mechanical work can be obtained from heat alone by an engine that repeats a regular 'cycle' of operations continuously?
- Is the efficiency of such an engine limited, and if so, how?
- Are other working substances preferable to steam for developing motive power from heat?

He felt that what was needed was a general theory that would apply to all possible heat engines, and he pointed out that, whatever the mechanism and whatever the working substance, work cannot be said to be produced *solely* from heat unless nothing but heat is supplied, and, at the end of the process, the working substance and all parts of the engine are in the same state as they were at the beginning.

He starts with two premises, both based on experience. Firstly, that perpetual motion is impossible. Secondly, that the production of 'motive power' solely from heat is possible only if there is a temperature difference and heat is transferred *by the engine* from a hotter body to a colder body. He points out that, for maximum effect, there should be as little *direct* transfer of heat between bodies at substantially different temperatures as is possible, since that would be a waste of heat that could be used by the engine to produce 'motive power'.

Having made these points he then introduces a simple ideal engine—now known as the Carnot engine—which consists of:

(1) a 'hot body' (at temperature T_h) large enough to supply a large amount of heat without a significant fall in temperature;
(2) a 'cold body' (at temperature T_c) large enough to absorb a large amount of heat without a significant rise in temperature;

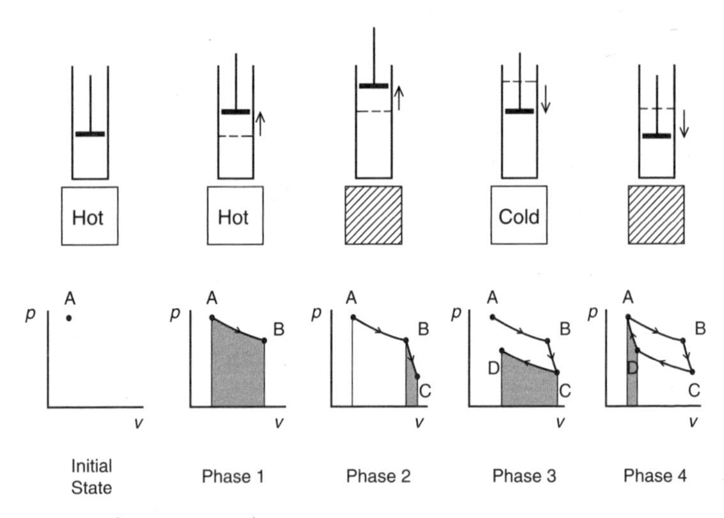

FIG 66 The Carnot cycle.

(3) a perfect heat insulator;
(4) a cylinder, fitted with a piston, enclosing an elastic fluid such as air.

It is assumed that the piston and the walls of the cylinder are perfect heat insulators, except for the bottom of the cylinder, which is a perfect conductor. Friction between the piston and the cylinder is assumed to be negligible. The load on the piston can be varied. If it just balances the upward pressure of the air in the cylinder, the piston will remain stationary. A slight increase in the load will force the piston downwards until the increased pressure of the air restores the balance. A slight decrease in the load will allow the piston to rise until the reduction in the pressure of the air restores the balance.

With the engine defined, he then considers what happens when he takes the elastic fluid—let's say air—through a cycle of operations that became known as the Carnot cycle—see Figure 66.

Let us look at the actual cycle, as it goes through four phases.

Phase 1 Imagine that the situation at the start corresponds to point A in Figure 66. The cylinder is standing on the hot body; the air in it is at the same temperature as the hot body; and the volume and pressure of the air are as shown in the figure. The load on the piston is reduced very gradually, so that the downward force on the piston is slightly less than the upward force caused by the pressure of the air in the cylinder; the piston moves slowly upwards as the air slowly expands. The work done by the air in phase 1 is equal to the shaded area under the line A–B.* A gas normally cools as it expands but, because the cylinder is standing on the hot body, cooling will be prevented by absorption of heat from that body. As the air increases in volume, its pressure falls slowly but the temperature remains the same as the hot body.

Phase 2 When the volume reaches a value corresponding to the (arbitrary) point B in Figure 66, the cylinder is removed from the hot body and placed on the perfect heat insulator. Again the load on the piston is reduced very gradually, but now as the air expands, pushing the piston upwards, the temperature of the air drops and the fall in pressure is steeper. The work done by the air on the piston is equal to the shaded area under the line B–C.

Phase 3 When the temperature of the air has fallen to that of the cold body (point C in Figure 66), the cylinder is removed from the perfect heat insulator and placed on the cold

* This is true because, for each small increase in volume, the work done is the product of the force exerted by the air on the piston and the distance the piston moves. But the force of the air on the piston is equal to the product of the pressure of the air in the cylinder and the area of the end of the piston (which is the same as the cross-sectional area of the cylinder). It follows that the work done during a small increase in volume is the product of the pressure of the air and the increase in its volume. For a larger increase in volume, the work done by the gas can be thought of as the sum of the products associated with a succession of small increases in volume.

body. The load on the piston is now *increased* very gradually, so the volume of the air is slowly reduced as the piston descends. The work done by the piston on the air is equal to the shaded area under the line C–D. Compression of a gas produces heat but the temperature of the air does not change because any heat produced is absorbed by the cold body.

Phase 4 When the volume of the air has fallen to a value corresponding to point D** in Figure 66, the cylinder is transferred again to the perfect heat insulator. Again, the load on the piston is *increased* very gradually, but now, as the piston descends and the air is compressed, the temperature of the air rises and the increase in pressure is steeper. When the temperature has reached that of the hot body, and the pressure and volume are back at their starting values, the cylinder is put back on the hot body and the cycle is complete. The work done by the piston on the air is equal to the shaded area under the line D–A.

The net work available from one cycle of the engine must be the difference between the work done by the expanding air on the piston in Phases 1 and 2, and the smaller amount of work done by the piston in compressing the air in Phases 3 and 4. This difference will be equal to the area enclosed by the curve ABCDA.

What makes the Carnot cycle so informative is a feature that is not immediately obvious from the above description. It is the cycle's reversibility. If at any moment during the very slow changes in any of the phases, the process were halted, the volume and pressure and temperature of the gas would scarcely change. At each moment during the cycle the gas is very nearly in an equilibrium state. It

** The position of point D is determined in a preliminary experiment. The cylinder filled with gas having the volume, pressure, and temperature corresponding to point A is placed on the perfect heat insulator and the load on the piston is very gradually reduced until the temperature has fallen to that of the cold body. The volume and pressure of the gas define point D.

follows that if at any moment the cause of the current change (i.e. the slight decrease or increase in the load on the piston) were reversed in sign, the change would occur in the opposite direction. That implies that the whole cycle could be run very slowly backwards, using mechanical work to transfer heat from the cold body to the air in the cylinder and from the air to the hot body. In other words, by running backwards the Carnot cycle can act as a refrigerator or a heat pump; and since friction is assumed to be negligible, the changes that occur when the cycle runs in one direction are the exact reverse of the changes that occur when it runs in the opposite direction.

Carnot realized that this led to a startling conclusion: that acting between two specified temperatures, no heat engine can be more efficient than an engine using the Carnot cycle. For suppose it could. Suppose the Carnot engine produces three units of work for a given amount of heat absorbed from the hot body, whereas the super-engine produces four units of work for the same amount of heat. If we couple a forward-running super-engine with a backwards-running Carnot engine, we can use three of the four units of work produced by the super-engine to drive the Carnot engine backwards restoring all the heat taken by the super-engine from the hot body. We shall then be left with one unit of available work though there has been no loss of heat from the hot body. This contradicts Carnot's second initial premise. If we use the available work to drive a machine, we shall need no fuel, and we shall have created a perpetual motion machine—which contradicts his first premise.

A further conclusion is that all heat engines using the Carnot cycle, and working between the same temperatures and with the same heat input, will produce the same amount of work, irrespective of the nature of the elastic fluid. Indeed the cycle need not involve the expansion of a gas. Any cycle made up of a sequence of reversible reactions such that heat is absorbed only at one temperature and expelled only at a lower temperature will behave in a similar way.

A thought experiment that is simple, ingenious, concise, and persuasive, and that leads to conclusions that are both startling and satisfying, deserves to be called elegant.

Unfortunately, in 1824 the work of Mayer and of Joule had not yet been done, and Carnot's understanding of his own cycle was limited by his belief (despite Rumford's cannon-boring experiments) in the caloric theory of heat. Since caloric was supposed to be indestructible, there could be no question of turning caloric into work. He thought (wrongly—and possibly partly influenced by his father's studies of hydraulic machinery) that deriving motive power from the transfer of heat from a hot body to a cold body was analogous to deriving motive power from the fall of water from one height to a lower height. The quantity of the water does not change as it falls, but the greater the change in height the greater the motive power of a given quantity of water.

After publication, Carnot's essay was favourably reviewed but then rarely mentioned over the next twenty years, apart from a paper by Clapeyron, a former fellow student with Carnot, who in 1834 wrote a paper containing a reformulation of Carnot's essay in a more conventional mathematical style. In 1832, Carnot was studying a cholera epidemic in Paris, and died of the disease. He was 36. Although most of his papers were destroyed, a bundle of twenty-three sheets of notes survived and were preserved by his brother Hippolyte Carnot. They suggest that Sadi Carnot had later rejected the caloric theory, accepted that heat was 'motion which has changed its form', and planned experiments to determine the mechanical equivalent of heat which anticipated some of those that would be done years later by Joule and by William Thomson (later Lord Kelvin). Unfortunately, these notes were not published until 1878.

In 1846 though, Thomson, having read Clapeyron's paper became interested in Carnot's approach, and in the following year, having heard Joule's talk at the Oxford meeting of the British Association for the Advancement of Science, he was persuaded that the caloric theory was no longer tenable, and that the interpretation of Carnot's findings needed to be changed. Only part of the heat absorbed from the hot body was transferred to the cold body. The rest was converted into work. A similar conclusion was reached independently by Rudolph Clausius, in Berlin, at about the same time.

The efficiency of a Carnot engine is the fraction of the heat absorbed from the hot body that is transformed into work. Since the heat transferred to the cold body is *not* transformed into work, it follows from the law of conservation of energy that the efficiency must be given by the expression

$$(Q_h - Q_c)/Q_h = 1 - Q_c/Q_h,$$

where Q_h is the amount of heat absorbed from the hot body, and Q_c is the amount of heat transferred to the cold body.

Since the efficiency of a Carnot engine depends only on the temperatures of the hot and cold bodies, and is not affected by the nature of the 'working substance', Thomson realized that by *defining* the ratio of the two temperatures (T_h/T_c) as the ratio Q_h/Q_c he would at last have a temperature-scale independent of the properties of any particular substance. Since no one could make a Carnot engine, this would not provide a practicable method for measuring temperature, but it would get over the difficulty that, because no two substances respond identically to changes in temperature, it is arbitrary to choose the behaviour of any one substance as the basis for a temperature scale. The best that had been done until this time was to base the scale on the behaviour of gases at very low pressure, when the differences in behaviour between different gases are very small. When such 'ideal gases' are heated, they all show almost the same linear increase in volume at constant pressure (or pressure at constant volume), and the graph extrapolates to zero volume (or zero pressure) at close to minus 273 °Celsius. At Joule's suggestion, Thomson correlated his Carnot scale with the ideal-gas scale by imagining the ideal gas as the working substance for the Carnot cycle, and defining the degree on the new scale by requiring that the difference in temperature between the melting point of ice and the boiling point of water be 100°. After Thomson became Lord Kelvin, the lowest possible temperature (close to minus 273° Celsius) began to be referred to as 0° Kelvin (0° K); the melting point of ice (0° Celsius) was close to 273° Kelvin; and the boiling point of water (100° Celsius) was close to 373° Kelvin.[†]

Both Thomson and Clausius produced versions of the second law of thermodynamics. Thomson's version was:

> It is impossible to derive a continuous supply of work by cooling a body to a temperature lower than that of the coldest of its surroundings.

Clausius' version was:

> Heat cannot of itself pass from a colder to a hotter body.

Each of these is derivable from the other, and they both make it clear that heat is available as a source of energy only if it can be transferred to a colder body.

Clausius went further, developing the important but difficult concept of entropy; but that is another story, and though it is an elegant one it needs more mathematics than is appropriate for this book.

[†] Curiously, since 1967 it has been the convention to omit the word 'degree' or the degree sign and refer to, say, the melting point of ice as 'close to 273 K'—or, in speech, as 'close to 273 Kelvins'.

NOTES

1 See W. Burkert (1962) trans. E.L. Minar, as *Lore and Science in Ancient Pythagoreanism* (1972) Harvard University Press.

2 *The Times*, 24 April 2007.

3 Thomas Digges, an English astronomer, was probably the first, in 1576.

4 In a monograph, *Exercitatio Anatomica de Motu Cordis et Sanguinis in Animalibus*, published in 1628 in Frankfurt am Main.

5 It was a chance observation by B. Erdmann and R. Dodge, who were studying the eye movements of people who were reading. See R. Dodge (1900) *Psychological Review*, 7:454–65.

6 If what moved at constant speed round the main orbit of a planet was not the planet itself but the centre of a small circle—they called it an epicycle—around whose circumference the planet moved at constant speed, the movement of the planet *as seen from a point on the Earth* would appear to vary in speed, and the planet would sometimes appear to move in the reverse direction. One could even envisage an epicycle on an epicycle.

A second fudge was to suppose that the centre of the main orbit was not the Earth itself but some point near it. This implied that the distance between the planet and the Earth varied, and a planet moving at constant speed round its orbit would appear to be moving faster when it was nearer the Earth.

A third fudge, and one even more open to criticism, was introduced by Ptolemy. Normally, the centre of an epicycle was supposed to move at constant speed along the circumference of the main orbit. Viewed from the centre of the main orbit it would, therefore, be moving with a constant angular velocity; viewed from a point at some distance from the centre, its angular velocity would vary.

What Ptolemy did was to choose a point at some distance from the centre, and suppose that the centre of the epicycle rotated with a constant angular velocity *around that point*. The speed of the centre as it moved along the circumference of the main orbit would therefore vary. This variation did, of course, represent a departure from the notion of constant speed round perfect circles, but the fact is that, by making use of these three fudges, Ptolemy was able to design a system containing thirty-nine (theoretical) 'wheels'—or forty if you include the outermost sphere with the fixed stars—which predicted the movements of all the heavenly bodies with remarkable accuracy.

7 See (i) article by Edward Rosen in the *Dictionary of Scientific Biography* (1970–80) Scribner's Sons, New York; (ii) Arthur Koestler (1979) *The Sleepwalkers*, Part 3, Hutchinson, London.

8 See (i) article by Owen Gingerich in the *Dictionary of Scientific Biography* (1970–80) Scribner's Sons, New York; (ii) Arthur Koestler (1979) *The Sleepwalkers*, Part 4, Hutchinson, London.

9 Quotation from Kepler's *Mysterium Cosmographicum* given in A. Koestler (1979) *The Sleepwalkers*, Part 4, Chapter 2, Section 1.

10 Quotation from Kepler's *Mysterium Cosmographicum* given in A. Koestler (1979) *The Sleepwalkers*, Part 4, Chapter 2, Section 1.

11 Quotation from Kepler's *Mysterium Cosmographicum* given in A. Koestler (1979) *The Sleepwalkers*, Part 4, Chapter 2, Section 2.

12 See (i) article by C. Doris Hellman in the *Dictionary of Scientific Biography* (1970–80) Scribner's Sons, New York; (ii) Koestler's *The Sleepwalkers*, Part 4, Chapters 4 and 5.

13 The *Alphonsine Tables* were improved versions of Ptolemaic tables, completed in 1252 in Toledo under the patronage of Alphonso X. The *Prutenic Tables* were based on Copernicus' observations and compiled by Erasmus Reinhold at the University of Wittenberg in 1551. Reinhold did not, however, accept Copernicus' heliocentric view.

14 Some estimates suggest that the total expenditure must have amounted to 5–10% of the gross national product of Denmark at that time. See Malcolm Longair (1984) *Theoretical Concepts in Physics*, p. 17, Cambridge University Press.

15 Letter to Michael Maestlin, 8 February 1601, trans. and quoted in Carola Baumgardt (1952) *Johannes Kepler: Life and Letters*, Gollancz, London.

16 Letter to Heydon, 1605, trans. and quoted in A. Koestler (1979) *The Sleepwalkers*, Part 4, Chapter 7, Section 1.

17 See M. Caspar (1959) *Kepler*, trans. and ed. C. Doris Hellman, Part III, Sections 3 and 4, Abelard-Schuman, London and New York.

18 *Ad Vittellionem Paralipomena, quibus Astronomiae Pars Optica traditur.* The book was remarkable, extending the work of the eleventh-century Arab Alhazen on the nature of light, the behaviour of lenses and mirrors, the pinhole camera, and the eye. Kepler showed the relation between the intensity of light and the distance from the source, and explained how refraction by the lens of the eye led to an inverted image on the retina. He didn't quite discover Snell's law of refraction, but he explained the use of lenses for people who were nearsighted or farsighted.

19 The quotation is from an article on Kepler by H. Dingle (1931) *Journal of the Royal Astronomical Society of Canada*, **XXV**(2):49–51.

20 A. Einstein (1952) Introduction to Baumgardt's *Johannes Kepler: Life and Letters*.

21 An English translation by W.H. Donahue, entitled *Johannes Kepler; New Astronomy*, was published by Cambridge University Press in 1992.

22 See A. Koestler (1979) *The Sleepwalkers*, Part 4, Chapter 6, Section 9.

23 Letter from Kepler to David Fabricius of 11 October 1605 in *Kepler's Gesammelte Werke* (1937–) ed. Walther von Dyck and Max Caspar, vol. 15, p. 241, Deutsche Forschungsgemeinschaft & Bayerischen Akademie der Wissenschaften, Munich.

24 See (i) article on Galileo Galilei by Stillman Drake in the *Dictionary of Scientific Biography*, (1970–80) Scribner's, New York. (ii) Stillman Drake (1978) *Galileo at Work: His Scientific Biography*, University of Chicago Press, Chicago and London.

25 The thought experiment is described in the article about Giovanni Benedetti by Stillman Drake in the *Dictionary of Scientific Biography*, (1970–80) Scribner's, New York.

26 Stillman Drake (1978) *Galileo at Work: His Scientific Biography*, Chapter 5, University of Chicago Press, Chicago and London.

27 When the ball reaches the end of the table it continues to move to the right with the same horizontal velocity v and it also begins the downward vertical motion associated with 'free fall'. Call the horizontal displacement x and the vertical displacement y.

At time t we can write:

$x = vt$

$y = kt^2$, where k is a constant.

Therefore, $y = kx^2/v^2$.

For any given horizontal velocity, k/v^2 will be constant. If we call that constant k' we can write

$y = k'x^2$,

which is the formula for a parabola.

(Because the horizontal displacement of the ball is always positive the actual curve will be half the parabola.)

28 Quoted in the article on Galileo by Agnes Mary Clerke in the 11th edition of the *Encyclopaedia Britannica*.

29 R.S. Westfall (1980) *Never at Rest: A Biography of Isaac Newton*, Cambridge University Press, Cambridge. For shorter accounts of Newton's life and work see S. Brodetsky (1927) *Sir Isaac Newton*, Methuen, London; and articles by: I.B. Cohen in the *Dictionary of Scientific Biography*, (1970–80) Scribner's Sons, New York; and by R.S. Westfall in the *Oxford Dictionary of National Biography*, (2004) Oxford University Press.

30 Conduitt's memorandum of a conversation with Newton, 31 August 1726 (Keynes MS 130.10 in King's College library), quoted in R.S. Westfall, (1980) *Never at Rest: A Biography of Isaac Newton*, Chapter 2, Cambridge University Press, Cambridge.

31 Years later Newton told John Conduitt, his niece's husband, that when he was examined for a scholarship in his third year at Trinity, he knew 'little or nothing' of Euclid's geometry. (See R.S. Westfall, (1980) *Never at Rest: A Biography of Isaac Newton*, p. 102, Cambridge University Press, Cambridge.)

32 I am grateful to Anson Cheung for pointing out the frequency of records (in Trinity College archives) of Newton's returns to Cambridge during this period.

33 The argument goes like this:

Newton showed that a body will move at a constant speed round a circular path if it is acted on by a force directed towards the centre of the circle, and the magnitude of the force is proportional to the square of the speed of the body, and inversely proportional to the radius of the circle. That is:

$$\text{Force} \propto v^2/r, \quad \text{(i)}$$

where v is the speed and r is the radius.

But Kepler's third law, obtained by comparing the orbital periods and distances from the Sun of the various planets, shows that for these planets the mean distances from the Sun and the orbital periods (and, therefore, the speeds) are not independent variables. The law is

$$r^3 \propto T^2,$$

where r is the mean distance from the sun and T is the orbital period.

Since, for an orbit that is close to circular

$$T = 2\pi r/v,$$

we have $r^3 \propto r^2/v^2$

or $v^2 \propto 1/r$ (ii)

Combining equations (i) and (ii), we get

$$\text{Force} \propto 1/r^2.$$

34 Quoted by R.G. Keesing (1998) *Contemporary Physics*, **39**:377–91.

35 Quoted by R.G. Keesing (1998) *Contemporary Physics*, **39**:377–91.

36 The radius of the moon's orbit had been worked out by Aristarchos (about 270 BC) and Hipparchus (about 129 BC) from studies of lunar and solar eclipses, using simple geometry.

37 Quoted by R.S. Westfall, (1980) *Never at Rest: A Biography of Isaac Newton*, p. 505, Cambridge University Press, Cambridge.

38 Quoted from *The Philosophy of Francis Bacon*, an address delivered at Cambridge on the occasion of the Bacon Tercentenary by C.D. Broad, 5 October 1926.

39 See: Sanborn C. Brown (1979) *Benjamin Thompson: Count Rumford*, MIT Press, Cambridge Mass.; G.I. Brown (1999) *Scientist, Soldier, Statesman, Spy: Count Rumford*, Sutton, Stroud; G.E. Ellis (1871) *Memoir of Sir Benjamin Thompson, Count Rumford*, Claxton, Remsen & Haffelfinger, Philadelphia; and articles in the *Dictionary of National Biography*, the *Oxford Dictionary of National Biography*, the *Dictionary of Scientific Biography* (1970–80) Scribner's New York, and the 11th edition of the *Encyclopaedia Britannica*.

40 Sanborn C. Brown, ed. (1968–70) *The Collected Works of Count Rumford*, vol. V, p. 5, Harvard University Press, Cambridge, Mass.

41 Benjamin Thompson (1792) *Philosophical Transactions of the Royal Society, London*, **82**:48–80.

42 See: 'The propagation of heat in Fluids' in Sanborn C. Brown, ed. (1968–70) *The Collected Works of Count Rumford*, vol. I, pp. 124–5, Harvard University Press, Cambridge, Mass.

43 In Munich, 2,600 out of a population of 60,000 were listed as beggars, or in need of public assistance, in one week. See 'On the Prevalence of Mendicity in Bavaria' in Sanborn C. Brown, ed. (1968–70) *The Collected Works of Count Rumford*, vol. V, pp. 13–18, Harvard University Press, Cambridge, Mass.

44 See 'An Account of an Establishment for the Poor at Munich' in Sanborn C. Brown, ed. (1968–70) *The Collected Works of Count Rumford*, vol. V, pp. 1–98, Harvard University Press, Cambridge, Mass.

45 See: *Of Chimney Fireplaces* in Sanborn C. Brown, ed. (1968–70) *The Collected works of Count Rumford*, vol. II, pp. 221–95, Harvard University Press, Cambridge, Mass.

46 Benjamin Thompson (1798) *Philosophical Transactions of the Royal Society, London* **88**:80–102.

47 I have modernized the units. Though most of us can cope with 'inches', 'pounds avoirdupois', and 'degrees Fahrenheit', few of us are familiar with 'grains troy'.

48 This figure is calculated assuming that the metal, which Rumford sometimes refers to as brass and sometimes as gunmetal, starts at 15.6 °C; that the melting point is 1000 °C (the figure for gunmetal) or 930 ° (the figure for brass); that the specific heat is 0.092 cal/(g °C); and that the latent heat of fusion is not too far from that of copper, which is 49 cal/g.

49 The quotation is from Rumford's pamphlet proposing the formation of what would become the Royal Institution.

50 See: W.H. Cropper (2001) *Great Physicists*, pp. 59–70, Oxford University Press, New York; D.S.L. Cardwell (1989) *James Joule: A Biography*, Manchester University Press, Manchester; articles in the *Dictionary of National Biography*; the *Oxford Dictionary of National Biography*; and the *Dictionary of Scientific Biography* (1970–80), Scribner's, New York.

51 J.P. Joule (1841) *Philosophical Magazine* series 3, **XIX**:260–77; J.P. Joule (1884–7) *Scientific Papers*, vol. 1, pp. 60–6, Physical Society of London.

52 To increase the conductivity of the water a small amount of sulphuric acid or potassium hydroxide was added; but because hydrogen ions (H^+) pick up an electron from the negative electrode much more readily than potassium ions (K^+), and hydroxyl ions (OH^-) lose an electron to the positive electrode much more readily than sulphate ions (SO_4^{2-}), the effect of the current is still to release molecules of hydrogen gas (H_2) at the negative electrode and molecules of both oxygen gas (O_2) and water at the positive electrode.

53 J.P. Joule (1843) *Memoirs of the Manchester Literary and Philosophical Society, 2nd Series*, **vii**:87. Also J.P. Joule (1884–7) *Scientific Papers*, vol. 1, pp. 109–23, Physical Society of London.

54 J.R. Mayer (1842) *Liebig's Annalen der Chemie und Pharmacie*, **42**:233. An English translation is included in R.B. Lindsay (1973) *Julius Robert Mayer: Prophet of Energy*, pp. 67–74, Pergamon, Oxford.

55 J.R. Mayer (1845) *Die organische Bewegung in ihrem Zusammenhang mit dem Stoffwechsel. (The Motions of Organisms and their Relation to Metabolism)*. Published privately after being rejected by *Liebig's Annalen*. An English translation is included in R.B. Lindsay (1973) *Julius Robert Mayer: Prophet of Energy*, pp. 75–145, Pergamon, Oxford.

56 J.P. Joule (1843) *Philosophical Magazine*, series 3, **XXIII**:263–76, 347–55, 435–43; J.P. Joule (1884–7) *Scientific Papers*, vol. 1, pp. 123–59.

57 J.P. Joule (1845) *Philosophical Magazine*, series 3, **XXVI**:369–83; J.P. Joule (1884–7) *Scientific Papers*, vol. 1, pp. 172–89.

58 J.P. Joule (1850) *Philosophical Transactions of the Royal Society* **140**:61–82; J.P. Joule (1884–7) *Joule Scientific Papers*, vol. 1, pp. 298–328.

59 C.F. Du Fay (1734) *Philosophical Transactions of the Royal Society*, **38**:258–66.

60 Letter from Benjamin Franklin in Philadelphia to Peter Collinson in London, 11 July 1747 in L. Jesse Lemisch, ed. (1961) *Benjamin Franklin: The Autobiography and Other Writings*, pp. 228–34, New American Library, New York.

61 Letter from Benjamin Franklin to Peter Collinson in London, 11 July 1747 in L. Jesse Lemisch, ed. (1961) *Benjamin Franklin: the Autobiography and Other Writings*, pp. 228–34, New American Library, New York.

62 Short accounts of Faraday's life and work can be found in the articles by Frank A.J.L. James in the *Oxford Dictionary of National Biography* (2004), Oxford University Press; by L. Pearce Williams in the *Dictionary of Scientific Biography*, (1970–80), Scribner's, New York; by John Tyndall in *The Dictionary of National Biography* (1908), vol. vi, Smith Elder, London; and also in Roy Porter, ed. (1994) *The Hutchinson Dictionary of Scientific Biography*, Helicon, Oxford. Fuller accounts are given by L. Pearce Williams (1965) *Michael Faraday*, Chapman & Hall, London; and by John Meurig Thomas (1991) *Michael Faraday and the Royal Institution*, Adam Hilger, Bristol. Faraday's papers about electricity are collected together in his *Experimental Researches in Electricity*, originally published in three volumes in 1839, 1844, and 1855, and republished in 1965 by Dover, New York.

63 It was founded in Scotland in the early eighteenth century by John Glas, formerly a Presbyterian minister, and modified by his son-in-law Robert Sandeman.

64 M. Rowbottom and C. Susskind (1984) *Electricity and Medicine: History of Their Interaction*, pp. 57–65. San Francisco Press, San Francisco, CA; H. Pixii (1832) see *Annales de chimie et de physique*, **50**:322–4.

65 (i) M. Faraday (1832) *Philosophical Transactions of the Royal Society*, 122:125–62. (ii) L. Pearce Williams (1965) *Michael Faraday*, Chapter 5, Chapman & Hall, London.

66 M. Faraday (1832) Bakerian Lecture, *Philosophical Transactions of the Royal Society*, 122:163–94, para. 159.

67 See L. Pearce Williams (1965) *Michael Faraday*, p. 196, Chapman & Hall, London.

68 W. Nicholson and A. Carlisle (1801) *Journal of Natural Philosophy, Chemistry and the Arts*, 4:179–87.

69 This would have been a better conductor of electricity than distilled water. They used water from the New River—an artificial watercourse built early in the seventeenth century to bring water to London from springs in Hertfordshire.

70 J. Meurig Thomas (1991) *Michael Faraday and the Royal Institution*, Chapter 4, Adam Hilger, Bristol.

71 Waltham Abbey is the name of the town in Essex that contains the abbey.

72 M. Faraday (1846) *Experimental Researches in Electricity*, 19th series.

73 The significance of the experiment is discussed on pages 133–4.

74 See article on Thomas Young in *Asimov's Biographical Encyclopedia of Science and Technology* (1972) Doubleday, Garden City, USA.

75 This account of Young's life is based largely on: (i) an autobiographical sketch written (in the third person) by Thomas Young two or three years before he died, and intended for publication in the *Encyclopaedia Britannica*, though never published. Although available to Hudson Gurney and to George Peacock, the sketch later disappeared and was rediscovered only in the 1970s in the archives of Francis Galton. It is printed, with an introduction by V.F. Hilts, as 'Thomas Young's "Autobiographical Sketch"' in *Proceedings of the American Philosophical Society* (1978) 122(4):248–60; (ii) George Peacock (1855) *Life of Thomas Young, M.D., F.R.S., &c.* John Murray, London; (iii) Alexander Wood (completed by Frank Oldham) (1954) *Thomas Young: Natural Philosopher* 1773–1829, Cambridge University Press; (iv) Andrew Robinson (2007) *The Last Man Who Knew Everything*, Oneworld,

Oxford. For shorter accounts, see articles by Geoffrey Cantor in the *Oxford Dictionary of National Biography* (2004), Oxford University Press; Edgar W. Morse in the *Dictionary of Scientific Biography* (1970–1980), Scribner's New York; and C.H. Lees (1900) *Dictionary of National Biography*, Smith Elder, London.

76 The Rosetta stone, carved in 196 BC, was found by Napoleon's soldiers in 1799 in an old wall in the village of el Rashid (Rosetta) on one of the branches of the Nile near the sea. The inscription is a decree passed by a council of priests affirming the royal cult of the 3-year-old Ptolemy V, but what makes it interesting is that the entire text is written three times: first in ancient Egyptian hieroglyphics, then in demotic script (developed in Egypt c. 650 BC), and finally in ancient Greek. Young was the first to notice a strong resemblance between some of the demotic characters and the more pictorial hieroglyphics, and he suggested that the demotic script was a mixture of imitations of hieroglyphics, which he believed to be purely symbolic (except when they were used for foreign names such as Ptolemy), and alphabetical letters. It was, however, his great rival the Frenchman Jean-François Champollion who made the startling discovery that hieroglyphics, too, could be used alphabetically (and not just for foreign names) as well as symbolically; and could therefore record the sound of the ancient Egyptian language.

77 See G. Peacock (1855) *Life of Thomas Young, M.D., F.R.S., &c.*, p. 31, John Murray, London.

78 H. Hingston Fox (1919) *Dr John Fothergill and his Friends: Chapters in Eighteenth Century Life*, p. 275, Macmillan, London.

79 Thomas had sent Brocklesby a copy, carefully written on vellum, of a Greek translation he had made of Wolsey's speech to Thomas Cromwell in Shakespeare's *Henry VIII*. Brocklesby had shown it to Burke, who expressed his appreciation by suggesting that Thomas should make a Greek translation of King Lear's 'horrid imprecations on his barbarous daughters'. Thomas did. (See G. Peacock (1855) *Life of Thomas Young, M.D., F.R.S., &c.*, pp. 19, 21, John Murray, London.).

80 The quotation is from Young's own account quoted in G. Peacock (1855) *Life of Thomas Young, M.D., F.R.S., &c.*, p. 89, John Murray, London.

81 *De corporis humani viribus conservatricibus.* Peacock (pp. 89–90) describes at some length the care Young took to use appropriate Latin.

82 *Dr Young's Reply to the Animadversions of the Edinburgh Reviewers, on Some Papers Published in the Philosophical Transactions*, produced as a pamphlet in 1804, and reproduced in G. Peacock (1855) *Miscellaneous works of the late Thomas Young, M.D., F.R.S., &c.*, vol. I. p. 199, John Murray, London.

83 Hudson Gurney (1831) *Memoir of the Life of Thomas Young*, J. & A. Arch, London.

84 T. Young (1807) *A Course of Lectures on Natural Philosophy and the Mechanical Arts*, 2 vols, Lecture XXXIII, Joseph Johnson, London.

85 Newton realized that the matter was not so simple. He pointed out that when a ray of light travelling in glass passes out of the glass into air (or into a vacuum), 'There is a Reflection as strong as in its passage out of Air into Glass, or rather a little stronger.' He concluded that, 'The Cause of Reflection is not the impinging of Light on the solid or impervious parts of Bodies, as is commonly believed.' See I. Newton (1704) *Optics*, 4th edn, reprinted 1931, Book II, Part III, Proposition VIII, pp. 262–3, G. Bell & Sons.

86 I. Newton (1730) *Opticks*, 4th edn, reprinted 1931, Book III, Part I, Query 17, pp. 347–8, G. Bell & Sons.

87 T. Young (1800) *Philosophical Transactions of the Royal Society*, **90**:106–50.

88 T. Young (1802) *Philosophical Transactions of the Royal Society*, **92**:12–48.

89 I. Newton (1730) *Opticks*, 4th edn, W. Innys, London.

90 T. Young (1804) *Philosophical Transactions of the Royal Society*, **94**:1–16.

91 T. Young, (1807) *A Course of Lectures on Natural Philosophy and the Mechanical Arts*, Lecture XXXIX.

92 The wording of Young's lecture *On the Nature of Light and Colours* suggests that the estimates were based on the two-slit experiments, but he starts the relevant paragraph with the words, 'From a comparison of various experiments,' and it is possible that they were based

on his earlier experiments using a slip of card to split the light beam.

93 Henry Brougham (1803) *Edinburgh Review*, 1:450–60; (1804) 5:97–103.

94 T. Young (i) (1813) *An Introduction to Medical Literature, Including a System of Practical Nosology*, W. Phillips, London; (ii) (1815) *A Practical and Historical Treatise on Consumptive Diseases*, Thomas Underwood and John Callow, London.

95 Quoted in A. Robinson (2006) *The Last Man Who Knew Everything*, pp. 173–4.

96 A. Fresnel (1821) *Annales de chimie et physique*, 17:101–12, 167–92, 312–16.

97 See letter by Fresnel quoted in A. Wood (1954) *Thomas Young*, pp. 198–200.

98 M. Faraday (1846) *Experimental Researches in Electricity*, 19th series.

99 J. Clerk Maxwell (1865) *Philosophical Transactions of the Royal Society* 155:459–512.

100 See M.S. Longair, (1984) *Theoretical Concepts in Physics*, p. 228, Cambridge University Press, Cambridge.

101 G.K. Batchelor, (1996) *The Life and Legacy of G.I. Taylor*, pp. 40–1, Cambridge University Press.

102 Extract from a talk given by Taylor at Rice University in 1963 and reproduced in G.K. Batchelor, (1996) *The Life and Legacy of G.I. Taylor*, Cambridge University Press.

103 G.I. Taylor, (1909) *Proceedings of the Cambridge Philosophical Society*, 15:114–15.

104 Quoted in an article on Erwin Schrödinger and Louis de Broglie in W.H. Cropper (2001) *Great Physicists*, Oxford University Press.

105 R. Feynman, (1965) *Character of Physical Law*, p. 129, M.I.T. Press, Cambridge, Mass.

106 See Chapter 4.

107 Descartes *Treatise on Man*.

108 Quotations from Borelli are taken from Sir Michael Foster (1901) *Lectures on the History of Physiology During the Sixteenth, Seventeenth and Eighteenth Centuries*, Cambridge University Press, London.

109 From Swammerdam, ed. Herman Boerhaave (1737–8) *Biblia Naturae*, vol. 3, Table XLIX, Figure viii, opposite p. 119. See also vol. 2, pp. 849–50.

110 C. Matteucci (1842) *Annales de chimie et de physique*, 3rd series, **6**:301–39.

111 C. Matteucci (1842) *Annales de chimie et de physique*, 3rd series, **6**:301–39.

112 C. Matteucci (1844) *Traité des phénomènes electro-physiologiques des animaux suivi d'études anatomiques sur le systhème nerveux et sur l'organe électrique de la torpille par Paul Savi*. Fortin, Masson et C'ie, Paris.

113 Quoted by Mary Brazier in her article 'The historical development of neurophysiology' in *American Handbook of Physiology* (1959), 1st edn, vol. 1, American Physiological Society, Washington DC.

114 H. Bence Jones, ed. (1852) *On Animal Electricity: Being an Abstract of the Discoveries of Emil Du Bois-Reymond*, pp. 182–3, Figures 46 and 47, John Churchill, London.

115 See H.E. Hoff and L.A. Geddes (1960) *Journal of the History of Medicine*, **15**:133–46.

116 Letter from Helmholtz to Du Bois-Reymond quoted (in English translation) in D. Cahan, ed. (1993) *Hermann von Helmholtz and the Foundations of Nineteenth-Century Science*, p. 83, University of California Press.

117 H. Helmholtz (1850) *Archiv für Anatomie und Physiologie, Leipzig*, 276–364.

118 J. Bernstein (1868) *Pflügers Archiv*, **1**:173–207.

119 L. Hermann (1872) See *Pflügers Archiv*, (1899) **75**:574–90.

120 J. Bernstein (1902) *Pflügers Archiv*, **92**:521–62.

121 F. Gotch and G.J. Burch (1899) *Journal of Physiology*, **24**:410–26.

122 E.D. Adrian, later Lord Adrian, was one of the first scientists to study the electrical changes associated with nerve impulses using thermionic valve amplifiers. He shared the 1932 Nobel Prize in Physiology or Medicine with Charles Sherrington.

123 E.D. Adrian (1928) *The Basis of Sensation*, Christophers, London.

124 K. Lucas (1909) *Journal of Physiology*, **38**:113–33.

125 A.L. Hodgkin (1939) *Journal of Physiology*, **94**:560–70; (1964) *The Conduction of the Nervous Impulse*, p. 35, Liverpool University Press. The account in the text is taken from I. Glynn (1999) *An Anatomy of Thought: The Origin and Machinery of the Mind*, Weidenfeld & Nicolson, London; Oxford University Press, New York.

126 A.L. Hodgkin and A.F. Huxley (1939) *Nature*, **144**:710–11.

127 K.S. Cole (1949) *Archives des sciences physiologiques*, **3**:253–8; G. Marmont (1949) *Journal of Cellular and Comparative Physiology*, **34**:351–82.

128 A.L. Hodgkin (1964) *The Conduction of the Nervous Impulse*, Chapter V, Liverpool University Press; A.L. Hodgkin (1992) *Chance and Design: Reminiscences of Science in Peace and War*, Chapter 31, Cambridge University Press, Cambridge.

129 From A.L. Hodgkin (1958) *Proceedings of the Royal Society, Series B*. **148**:1–37; A.L. Hodgkin and A.F. Huxley (1952) *Journal of Physiology, London*, **116**:449–72 and 473–96.

130 R.D. Keynes (1948) *Journal of Physiology*, **107**:35P; R.D. Keynes (1951) *Journal of Physiology*, **114**:119–50.

131 It was awarded jointly to Hodgkin, Huxley, and John Eccles, but Eccles' share in the prize was awarded for his recordings of electrical changes in nerve cell bodies in the spinal cords of mammals during reflex activity.

132 The accounts in this chapter are largely based on earlier accounts in I. Glynn (1999) *An Anatomy of Thought, the Origin and Machinery of the Mind*. Weidenfeld & Nicolson, London; Oxford University Press, New York.

133 H. von Helmholtz (1856–76) *Treatise on Physiological Optics*, vol. 3, Chapter 26; Voss, Hamburg and Leipzig; English translation 1925, reprinted Dover, New York, 1962.

134 From V.S. Ramachandran (1988) *Nature*, **331**:163–6.

135 M.A. Goodale, A.D. Milner, L.S. Jakobson, and D.P. Carey (1991) *Nature*, **349**:154–6: A.D. Milner, D.I. Perrett, R.S. Johnston, *et al.* (1991) *Brain*, **114**:405–28.

136 From N. Geschwind (1979) *Scientific American*, **241** (September): 158–68.

137 M.J. Dejerine (1892) *Compte Rendu Memoire-Société Biologie*, **4**:61–90.

138 J.H. Trescher and F.R. Ford (1937) *Archives of Neurology and Psychiatry*, **37**:959–73.

139 N. Geschwind, F.A. Quadfasel, and J.M. Segarra (1968) *Neuropsychologia*, **6**:327–40.

140 D.M. Hunt, K.S. Dulai, J.K. Bowmaker, and J.D. Mollon (1995) *Science*, **267**:984–8.

141 The passage is from Helmholtz's *Handbuch der Physiologischen Optik*, quoted in translation in S. Zeki (1993) *A Vision of the Brain*, p. 54, Blackwell, Oxford.

142 E.H. Land (1959) *Proceedings of the National Academy of Sciences of the USA*, **45**:115–29.

143 E.H. Land (1986) *Proceedings of the National Academy of Sciences of the USA*, **83**:3078–80.

144 Remarkably, a similar suggestion was made in 1789, by the French geometer Gaspard Monge, who demonstrated experiments that led him to it in a talk in the Académie Royale des Sciences two weeks before the French Revolution, and twelve years before Young produced his trichromatic theory. See article by John Mollon in T. Lamb and J. Bourriau, eds (1995) *Colour: Art and Science*, pp. 127–50, Cambridge University Press.

145 E.H. Land (1983) *Proceedings of the National Academy of Sciences of the USA*, **80**:5163–9.

146 S. Zeki (1983) *Neuroscience* **9**:741–65 and 767–81.

147 M. Ridley (1999) *Genome: The Autobiography of a Species in 23 Chapters*, p. 12, Fourth Estate, London.

148 There are some cells, such as mature red blood cells, that lack nuclei.

149 The phrase 'survival of the fittest' was not used by Darwin in *The Origin of Species* (published 1859) but was introduced by Herbert Spencer.

150 This account of Mendel and his work is largely based on (i) H. Iltis (1932) *Life of Mendel*, Allen & Unwin, London (trans. E. and C. Paul from the 1924 German original, Springer, Berlin); (ii) V. Orel (1996) *Gregor Mendel: The First Geneticist*, (trans. E. Finn), Oxford University Press; (iii) R.M. Henig (2000) *A Monk and Two Peas*, Weidenfeld & Nicolson, London; (iv) R. Olby (1985) *Origins of Mendelism* 2nd edn, University of Chicago Press.

151 G. Mendel (1866) 'Versuche über Pflanzenhybriden', *Verhandlungen des Naturforschenden Vereins in Brünn*, **4**:3–47.

152 I am indebted to Robin Henig's *A Monk and Two Peas* for the details of Mendel's technique for cross-pollination.

153 W. Bateson (1902) *Mendel's Principles of Heredity. A Defence*, Cambridge University Press, London.

154 This account of Morgan and his work is largely based on G. Allen (1978) *Thomas Hunt Morgan: The Man and his Science*, Princeton University Press.

155 C.E. McClung (1902) *Biological Bulletin*, **3**:43–84.

156 Th. Boveri, (1902) *Verhandlungen der Physikalisch-Medizinischen Gesellschaft zu Würzburg*, 35:67–90; W.S. Sutton (1903) *Biological Bulletin*, **4**:231–51.

157 T.H. Morgan (1916) *A Critique of the Theory of Evolution*, pp. 118–37, Princeton University Press. See also F.A. Jannsens (1909) *La Cellule*, **25**:389–411.

158 G. Allen (1978) *Thomas Hunt Morgan: The Man and His Science*, pp. 166–7, Princeton University Press.

159 See T.H. Morgan (1993) Croonian Lecture of 1922, *Proceedings of the Royal Society*, **94**:162–97.

160 O.T. Avery, C.M. MacLeod, and M. McCarty (1944) *Journal of Experimental Medicine*, **79**:137–58.

161 This account of Avery and his work is based partly on individual papers and partly on (i) R.J. Dubos (1956) *Biographical Memoirs of the Royal Society*, **2**:35–43; (ii) R.J. Dubos (1976) *The Professor, the Institute and DNA*, Rockefeller University Press, New York; (iii) H.F. Judson (1979) *The Eighth Day of Creation: The Makers of the Revolution in Biology*,

pp. 33–41, Jonathan Cape, London; (iv) Article on Oswald T. Avery, by Alan S. Kay in the *Dictionary of Scientific Biography* (1970) Scribner, New York; (v) The Oswald T. Avery Collection (*Shifting Focus: Early Work on Bacterial Transformation*, 1928–1940) US National Library of Medicine. Accessed from the Internet, September 2007.

162 F. Griffith (1928) *Journal of Hygiene*, **27**:113–59.

163 J.L. Alloway (1932) *Journal of Experimental Medicine*, **55**:91–9; J.L. Alloway (1933) *Journal of Experimental Medicine*, **57**:265–78.

164 In a letter from Oswald Avery to his brother Roy, written in May 1943 (from R.D. Hotchkiss (1966) in *Phage and the Origins of Molecular Biology*, eds J. Cairns, G.S. Stent, and J.D. Watson, pp. 185–6, Cold Spring Harbor Laboratory).

165 A. Klug (2004) *Journal of Molecular Biology*, **335**:3–26.

166 A. Klug (2004) *Journal of Molecular Biology*, **335**:3–26.

167 E. Chargaff (1950) *Experientia*, **6**:201–9.

168 R. Olby (1994) *The Path to the Double Helix*, Chapter 21, Dover, New York.

169 Chapter 6.

170 See article on Max von Laue by Armin Hermann in the *Dictionary of Scientific Biography* (1970–80) Scribner's, New York.

171 There were three papers: J.D. Watson and F.H.C. Crick (1953) *Nature*, **171**:737–8; M.H.F. Wilkins, A.R. Stokes, and H.R. Wilson (1953) *Nature*, 171:738–40; R.E. Franklin and R.G. Gosling (1953) *Nature*, **171**:740–1.

172 R. Olby (1994) *The Path to the Double Helix*, Dover, New York.

173 H.F. Judson (1979) *The Eighth Day of Creation*, Jonathan Cape, London.

174 A. Klug (2004) *Journal of Molecular Biology*, **335**:3–26.

175 M. Meselson and W.F. Stahl (1958) *Proceedings of the National Academy of Sciences, USA*, **44**:671–82.

176 By John Cairns, a former director of the *Cold Spring Harbor Laboratory*—see H.F. Judson (1979) *The Eighth Day of Creation*, p. 188, Jonathan Cape, London.

177 F.H.C. Crick, J.S. Griffith, and L.E. Orgel, (1957) *Proceedings of the National Academy of Sciences*, USA, **43**:416–21.

178 Francis Crick, (1988) *What Mad Pursuit*, pp. 99–100, Basic Books, New York.

179 M.B. Hoagland (1959) *Scientific American* **201**(December):55–61.

180 See P. Lipton (2004) *Inference to the Best Explanation*, 2nd edn, Chapter 4, Routledge, Abingdon.

181 See: W.H. Cropper (2001) *Great Physicists*, pp. 43–50, Oxford University Press, New York; article by J.F. Challey in the *Dictionary of Scientific Biography* (1970–80) Scribner's, New York; and articles on heat and on thermodynamics in the 11th edition of the *Encyclopaedia Britannica*.

182 The quotation is from W.H. Cropper (2001) *Great Physicists*, p. 50, Oxford University Press, New York.

183 *Réflexions sur la puissance motrice du feu et sur les machines propres à développer cette puissance.*

INDEX

Bold numbers refer to Figures. *Endnotes* are indicated by 'n.' followed by the number of the endnote. *Footnotes* are indicated by 'fn' preceded by the number of the page on which the footnote appears.